RECENT ADVANCES in
COMBUSTION MODELLING

Series on Advances in
Mathematics for Applied Sciences - Vol. 6

RECENT ADVANCES IN COMBUSTION MODELLING

Editor

B. Larrouturou

CERMICS, INRIA
Sophia-Antipolis
France

World Scientific
Singapore • New Jersey • London • Hong Kong

Published by

World Scientific Publishing Co. Pte. Ltd.
P O Box 128, Farrer Road, Singapore 9128
USA office: 687 Hartwell Street, Teaneck, NJ 07666
UK office: 73 Lynton Mead, Totteridge, London N20 8DH

RECENT ADVANCES IN COMBUSTION MODELLING

ISBN 981-02-0380-2

Printed in Singapore by JBW Printers & Binders Pte. Ltd.

Foreword

This volume gathers the written contributions of the courses and lectures given during the session on combustion modelling, which was held in INRIA in Rocquencourt (France) in March 1989, in the framework of the session series "Problèmes Non Linéaires Appliqués" (Nonlinear Applied Problems) organised under the auspices of CEA (Commissariat à l'Energie Atomique), EDF (Electricité De France) and INRIA (Institut National de Recherche en Informatique et Automatique).

The program of this combustion session consisted of two main courses, "Introduction to Combustion Modelling" by Prof. N. Peters (RWTH Aachen), and "Turbulent Combustion" by Prof. F. A. Williams (University of California, San Diego), and of a seris of lectures by J. Buckmaster (University of Illinois), S. Candel (Ecole Centrale de Paris), P. Clavin (Université de Provence, Marseille), A. Liñan (ETS, Madrid), X. Lys (Renault, Paris), J. Warnatz (Universität Stuttgart) and the editor. The session was attended by 65 participants, either students, researchers or engineers, coming from various European countries, and mainly from France.

As shown by the contributions in this volume, the combustion science has made in the last five years several important progresses, which are of interest from both the scientific and industrial points of views. Also, there has been in the same period an increasing number of interactions between scientists interested in combustion and engineers concerned by combustion devices for vehicle propulsion or energy production. This is why it is hoped that this volume, where some of the worldwide experts of the combustion community present in a pedagogical way some recent advances in combustion modelling, can help for further progress in both combustion science and technology.

I wish to thank here the three institutions who decided to hold this combustion session (CEA, EDF and INRIA), and all persons who helped me in organising this session. In particular, it has been a pleasure to co-organise the session with Henri Berestycki and Pierre-Louis Lions, and I am also very grateful to Valérie Bruker and Agnès Couaillier from the INRIA Public Relations Department, whose help has been crucial for the success of the session.

Bernard Larrouturou
Editor

Table of Contents

Foreword v

Modeling of Microgravity Ignition Experiments 1
J. Buckmaster, C. J. Lee, G. Joulin, & P. Ronney

Coherent Flamelet Model: Applications and Recent Extensions 19
S. Candel, D. Veynante, F. Lacas, E. Maistret, N. Darabiha, & T. Poinsot

Modélisation Physique, Numérique et Mathématique des Phénomènes de Propagation de Flammes 65
B. Larrouturou

Introduction à L'étude de la Combustion 121
N. Peters & C. Kennel

Simulation of Ignition Processes 185
J. Warnatz

Turbulent Combustion 221
F. A. Williams

MODELING OF MICROGRAVITY IGNITION EXPERIMENTS

J. Buckmaster, C. J. Lee, University of Illinois,
G. Joulin, CNRS Poitiers,
& P. Ronney, Princeton University

ABSTRACT

We present a review of recent theoretical results relevant to ignition experiments in the NASA-Lewis 2.2 sec drop tower. Two topics are discussed: the $\sqrt{t}$ radial dependence of unsteady spherical flames initiated at a point; and flame-balls, stationary spherical premixed flames that arise from Turing instabilities of unsteady flames when the Lewis number is small.

INTRODUCTION

In recent years a large number of μg ignition experiments have been carried out in the NASA-Lewis 2.2 sec drop tower which have revealed a variety of hitherto unknown phenomena, particularly for mixtures that have a characteristic Lewis number that is less than unity[1] - [7]. The analysis of simple model equations can provide insight into some of these effects [8] - [10], and here we provide a brief review of the theoretical results and the experimental work that motivates them. In addition some new results on flame-ball stability (see below) are presented. Two distinct phenomena are discussed: the $\sqrt{t}$ behavior of self-extinguishing flames [4], [9]; and flame-balls [1], [10].

Self-extinguishing flames (SEFs) are observed in a variety of gases, e.g. lean methane-air, lean ammonia-air, and rich propane-air mixtures. Fig. 1 is a schematic which reveals some of their essential characteristics. Curve A is the μg minimum ignition energy curve; for initiation energies above this curve the flame that is generated propagates through the entire mixture to the containment walls. Below this curve only partial propagation occurs, and B and C are curves of fixed extinction radius corresponding to SEFs. The radius of a SEF grows like $\sqrt{t}$ where t is the time from initiation.

The history of these flames is remarkably sensitive to the initial conditions. A point such as P_2 can lead to a flame that extinguishes after releasing energy one thousand times greater than the spark energy and one might naturally expect that this would overwhelm the details of initiation. And yet a modest (< 100%) increase in spark energy to P_1 leads to a flame that propagates indefinitely. The large amount of energy released by an SEF clearly distinguishes it from a classical non-ignition in which the released energy is at most one order of magnitude greater than the spark energy.

SEFs are not observed in mixtures that have small Lewis Numbers (e.g. lean hydrogen-air). Turing instabilities[11] cause the spherical propagating flame to break up and form a strongly three-dimensional cellular structure. For near-limit mixtures, individual cells tend to form independent spherical flames that appear to have stationary structure. These are flame-balls, they have a 1g non-spherical counterpart [12], and their stability needs explanation, for one would expect them to act as ignition sources for the surrounding mixture.

RADIAL PROPAGATION AND $\sqrt{t}$ BEHAVIOR

SEFs are observed for mixtures with Lewis numbers $0.5 \lesssim Le \lesssim 1$; for Le < .5 cellular structures occur. It is therefore natural to ask what radial history can be expected of an unsteady spherical flame when Le < 1. This problem has been examined in [8], [9] using asymptotic methods applied to simple model equations with the restriction that Le is bounded away from and below 1. For limit methane mixtures $Le \simeq 0.96$ [5], undoubtedly too close to 1 for the theory to be applicable, so that it is important to emphasize that although the theory yields $\sqrt{t}$ behavior there must be more to the physics than is yet dreamt of in the model. This *caveat* is reinforced by the observation that the theory does not explain the extraordinary sensitivity to the spark energy that is sometimes observed.

Reference [9] is concerned with the model equations

$$\rho C_p \frac{\partial T}{\partial t} = \frac{\lambda}{r^2}\frac{\partial}{\partial r}\left(r^2 \frac{\partial T}{\partial r}\right) + QBe^{-E/2RT_*}\,\delta(r - r_*)\,, \tag{1}$$

$$\rho \frac{\partial Y}{\partial t} = \frac{\rho D}{r^2}\frac{\partial}{\partial r}\left(r^2 \frac{\partial Y}{\partial r}\right) - B\,e^{-E/2RT_*}\,\delta(r - r_*)\,,$$

$$\rho = \text{constant},$$

$$Y = 0 \text{ for } r < r_*; \quad Y \to Y_\infty,\ T \to T_\infty \text{ as } r \to \infty.$$

The constant density model is not adopted in [8] so that (lc) is replaced by the continuity equation and advection terms are added to (1a, b), but this refinement has no qualitative impact on the evolution of the flame-sheet.

Asymptotic solutions of these equations can be constructed valid in the limit $\theta = EC_p/RQ \to \infty$, and take one of three forms depending on the radius of the flame, and therefore the time from initiation.

We introduce non-dimensional variables

$$\bar{r} = \frac{r}{R_{ref}},\ \bar{t} = \frac{\lambda t}{\rho C_p R_{ref}^2},\quad \bar{T} = \frac{C_p T}{Q},\quad \bar{B} = B\frac{R_{ref} C_p}{\lambda},\quad \theta = \frac{EC_p}{RQ} \tag{2}$$

where $$R_{ref} = \frac{Y_\infty \lambda}{LeBC_p}\exp\left[\frac{E}{2RT_{ref}}\right] \tag{3}$$

and T_{ref} is not yet assigned. Then

$$\frac{\partial \bar{T}}{\partial \bar{t}} = \frac{1}{\bar{r}^2} \frac{\partial}{\partial \bar{r}} (\bar{r}^2 \frac{\partial \bar{T}}{\partial \bar{r}}) + \bar{B}\, e^{-\theta/2\bar{T}_*}\, \delta(\bar{r} - \bar{r}_*) \tag{4}$$

$$\frac{\partial Y}{\partial \bar{t}} = \frac{1}{Le} \cdot \frac{1}{\bar{r}^2} \frac{\partial}{\partial \bar{r}} (\bar{r}^2 \frac{\partial Y}{\partial \bar{r}}) - \bar{B}\, e^{-\theta/2\bar{T}_*}\, \delta(\bar{r} - \bar{r}_*)\ .$$

The smallest flames of the asymptotic triad are called Zeldovich flames. We make the choice

$$T_{ref} = T_b \equiv T_\infty + \frac{QY_\infty}{C_p Le} \tag{5}$$

(then $R_{ref} = R_z$ the Zeldovich radius), and seek solutions that evolve on the slow time scale $\tau = 0(1)$ where

$$\bar{t} = \theta^2 \tau\ . \tag{6}$$

The near-field ($\bar{r} = 0(1)$) is quasi steady to second order in $1/\theta$ so that solutions are easily constructed; far-field ($\bar{r} = 0(\theta)$) perturbations to $\bar{T}_\infty$, Y_∞ satisfy the spherical heat equation which can be solved using the Laplace transform. In this way an evolution equation for $\bar{r}_*(\tau)$ can be derived whose precise form depends on the initial conditions for the far-field problem. The simplest example of such an equation is

$$\frac{Y_\infty(\sqrt{Le} - 1)}{2\sqrt{\pi}\, \bar{T}_b^2\, Le} \int_0^\tau as\ \frac{\dot{\bar{r}}_*(s)}{\sqrt{\tau - s}} + \ln \bar{r}_* = 0 \tag{7}$$

which has the late-time solution

$$\bar{r}_* \sim \frac{2\, \bar{T}_b^2\, Le}{\sqrt{\pi}\, Y_\infty (1 - \sqrt{Le})} \sqrt{\tau} \ln \tau \quad as \quad \tau \to \infty\ , \tag{8}$$

and this will be the asymptotic behavior for all such equations. These flames are hot ($T_* \sim T_b$), small ($r_* \sim R_z$) and fast,

$$\frac{dr_*}{dt} \sim \frac{1}{\theta^2} \frac{M_{ad}(T_b)}{\rho}$$

where
$$M_{ad}(T_b) = \frac{B}{Y_\infty} \exp\left(\frac{-E}{2RT_b}\right)$$

so that $M_{ad}(T_b)/\rho$ is the adiabatic flame speed evaluated at the temperature T_b. Note that T_b is larger than the adiabatic flame temperature $T_a = T_\infty + QY_\infty/C_p$.

As r_* becomes large compared to R_z, T_* decreases significantly below T_b signalling the start of a different region requiring a different asymptotic treatment. Retaining the choice (5) we seek solutions of (4) in terms of the new variables

$$\eta = \bar{r}/\sqrt{\bar{t}}, \quad \sigma = \frac{1}{\theta} \ln \bar{t} \tag{9}$$

so that

$$\frac{1}{\theta}\frac{\partial \bar{T}}{\partial \sigma} = \frac{1}{2}\eta\frac{\partial \bar{T}}{\partial \eta} + \left[\frac{\partial^2 \bar{T}}{\partial \eta} + \frac{2}{\eta}\frac{\partial \bar{T}}{\partial \eta}\right] - \bar{B}\, e^{\theta\sigma/2 - \theta/2\bar{T}_*}\,\delta(\eta - \eta_*)\,,$$

$$\frac{1}{\theta}\frac{\partial \bar{T}}{\partial \sigma} = \frac{1}{2}\eta\frac{\partial \bar{T}}{\partial \eta} + \left[\frac{\partial^2 \bar{T}}{\partial \eta} + \frac{2}{\eta}\frac{\partial \bar{T}}{\partial \eta}\right] - \bar{B}\, e^{\theta\sigma/2 - \theta/2\bar{T}_*}\,\delta(\eta - \eta_*)\,. \tag{10}$$

To first order σ derivatives may be neglected so that it is only necessary to solve ordinary differential equations. In this way it can be shown that the flame temperature is

$$\bar{T}_* = \frac{\bar{T}_b}{1 + \sigma\bar{T}_b}, \tag{11}$$

and the flame radius is determined by

$$\frac{Le}{Y_\infty}(\overline{T}_* - \overline{T}_\infty) = \frac{I(1, \eta_*)}{I(Le, \eta_*)} e^{\eta_*^2 (1 - Le)/4} , \tag{12}$$

where
$$I(a,x) \equiv \int_\infty^x \frac{d\eta}{\eta^2} e^{-a\eta^2/4} .$$

Note that
$$\overline{r}_* = \overline{t}^{1/2} \eta_* \left(\frac{1}{\theta} \ln \overline{t}\right). \tag{13}$$

This solution is valid for

$$0 < \sigma < \sigma_c \equiv \frac{Y_\infty(Le^{-1} - 1)}{\overline{T}_a \overline{T}_b} \tag{14}$$

with η_* ranging from 0 to ∞, T_* from T_b to T_a. These flames are much larger than R_z ($r_* \sim R_z e^{\theta\sigma/2}$) and are cooler and slower than Zeldovich flames.

As $\sigma \to \sigma_c (\eta^* \to \infty)$ a third and final asymptotic analysis must be initiated. This corresponds to what are called Slowly Varying Flames (SVFs) 13].

We make the choice

$$T_{ref} = T_a$$

and seek a solution in terms of the variables

$$x = \overline{r} - \overline{r}_* , \quad \tau = \overline{t}/\theta, \quad R_*(\tau) = \frac{1}{\theta} \overline{r}_* . \tag{15}$$

so that, with $0(1/\theta^2)$ terms omitted,

$$\frac{1}{\theta} \frac{\partial \overline{T}}{\partial \tau} - \dot{R}_* \frac{\partial \overline{T}}{\partial x} = [\frac{\partial^2 \overline{T}}{\partial x^2} + \frac{2}{\theta R_*} \frac{\partial \overline{T}}{\partial x}] + \overline{B} e^{-\theta/2\overline{T}_*} \delta(x), \tag{16}$$

$$\frac{1}{\theta}\frac{\partial Y}{\partial \tau} - \dot{R}_* \frac{\partial Y}{\partial x} = \frac{1}{Le}\left[\frac{\partial^2 Y}{\partial x^2} + \frac{2}{\theta R_*}\frac{\partial Y}{\partial x}\right] - \overline{B}\, e^{-\theta/2\overline{T}_*}\, \delta(x).$$

Solution yields the evolution equation

$$\frac{\ddot{R}_*}{\dot{R}_*^2} + \frac{2\overline{T}_a^2}{Y_\infty(Le^{-1} - 1)} \dot{R}_* \ln (Le\, \dot{R}_*) = \frac{2}{R_*} \tag{17}$$

which can be transformed to

$$\frac{dV}{dL} + 2V^2 \ln V = \frac{2V}{L} \tag{18}$$

where $$V = Le\dot{R}_*\,,\ L = \frac{\overline{T}_a^2\, Y_\infty\, R_*}{Y_\infty^2(1 - Le)}. \tag{19}$$

As $L \to 0$, $V \to \infty$ (to match with the solution on the σ scale)

$$V \sim \frac{1}{L}\left[-\frac{3}{2}\cdot\frac{1}{\ln L} + 0\left(\frac{\ln(-\ln L)}{\ln^2 L}\right)\right] \tag{20}$$

so that, within logarithmic factors, we once again have $\sqrt{t}$ behavior. As $L \to \infty$, we have $V \to 1$ and we recapture the plane adiabatic flame.

These analyses reveal that for large Zeldovich flames, small SVFs, and all intermediate flames, the flame radius behaves like $\sqrt{t}$ to within logarithmic factors (equations (8), (13), (20)), suggesting a connection with SEFs. Extinguishment does not occur, but can be generated by the addition of volumetric heat losses to the model [8], [9]. Earlier we noted some areas in which the theoretical results differ from the experimental facts -- additional discussion may be found in [9].

FLAME BALLS WHEN Le < 1

When Le < .5 cellular structures are observed in the μg ignition experiments, and for near-limit mixtures the cells separate and close to form what appear to be stationary spherical flames. With only 2.2 sec of test time there is some uncertainty in this connection, but the evidence justifies a theoretical treatment [10].

The starting point is the observation first made by Zeldovich that equations (1) have a stationary solution:

$$r_* = R_z$$

r < r* $$Y = 0,\ T = T_b \tag{21}$$

r > r* $$Y = Y_\infty\left(1 - \frac{r_*}{r}\right),\ T = T_\infty + \frac{QY_\infty}{C_p\,Le}\cdot\frac{r_*}{r}\,.$$

This is an unusual type of premixed flame for the fluxes are solely diffusional with the mixture flux directed inwards from infinity, the products and heat fluxes directed outwards. However, these flames are unstable and, if disturbed, will either propagate outwards or collapse. Certainly the possibility of outward propagation is easy to understand intuitively -- the hot flame acts as an ignition source for the surrounding mixture. It follows that the existence of stationary flames requires a stabilizing mechanism, and in reference [10] it is proposed that volumetric heat losses can play this role.

Equation (1a) is now replaced by

$$\rho\, C_p \frac{\partial T}{\partial t} = \frac{\lambda}{r^2}\frac{\partial}{\partial r}\left(r^2 \frac{\partial T}{\partial r}\right) + QB\, e^{-E/2RT_*}\,\delta(r - r_*) - f(T) \tag{22}$$

where f(T) is the heat loss term. Choosing the size of this term presents a small technical difficulty. One choice is to pick f so that the coefficient of the non-dimensional term (using equations (2) with $R_{ref} = R_z$) is

$O(1/\theta^2)$. Then in the far-field ($\bar{r} = O(\theta)$) where δT is $O(1/\theta)$, this term, like the diffusion term, is $O(1/\theta^3)$ and has a significant effect on the solution. However, in the near-field ($\bar{r} = O(1)$) it is $O(1/\theta^2)$ and so plays no important role, which is physically unsatisfactory in view of the high temperatures that prevail there. Another choice is to assume that f is identically zero in the far-field and the nondimensional term is $O(1/\theta)$ so that near-field losses have a significant effect. This is the choice made in [10], specifically with f identically zero beyond the flame-sheet. Yet this also discards some physics, for the impact on the stability is different for far-field and near-field losses. Reference [14] combines the two, and we shall remark on the consequences a little later, but first we will summarize the main results from [10].

Now the stationary radius R_0 differs from R_z:

$$1 = \frac{R_0}{R_z} e^{-q R_z^2 (R_0/R_z)^2} \qquad (23)$$

where q is a volume integral of the function f (whose exact temperature dependence is unimportant), a measure of the total heat loss; this solution is shown in Fig. 2. If the heat losses are too large a stationary solution does not exist, and for heat losses smaller than the quenching value there are two solutions, a small flame and a large flame. The radius of the small flame approaches R_z as $q \to 0$, that of the large flame approaches ∞.

The one-dimensional stability of these solutions can be examined as in the $\sqrt{t}$ problem by constructing an evolution equation that is a generalization of (7) i.e.

$$\frac{Y_\infty(\sqrt{Le} - 1)}{2\sqrt{\pi}\, \bar{T}_b^2\, Le} \int_0^\tau ds \frac{\dot{\bar{r}}_*(s)}{\sqrt{\tau - s}} = \ln\left[\frac{R_z}{R_0\, \bar{r}_*}\right] + \frac{\bar{r}_*^2 F(\bar{T}_b)}{6\bar{T}_b^2} \qquad (24)$$

where non-dimensional variables are determined by (2) with $R_{ref} = R_0$, and

$$F = \frac{\theta R_0{}^2 C_p f}{\lambda Q} \ .$$

This can be used to show that the entire small-flame branch is unstable (Fig. 2).

A three-dimensional stability analysis can be carried out by examining the response to infinitesimal perturbations characterized by flame-sheet perturbations of the form

$$r_* = 1 + \delta \, e^{im\phi} \, P_n^m (\cos \theta) \, e^{\alpha t} \qquad (25)$$

where $(r, \theta \ \phi)$ are spherical coordinates, and the $\{P_n^m\}$ are spherical harmonics of the first kind. Steady solutions for which $R_0 > e^{5/2} R_z$ are unstable (Fig. 2).

When far-field heat losses are incorporated, the stability picture is modified in an important way. Now the band of solutions unstable to one-dimensional disturbances wraps around the nose of the response curve, with the neutral stability point lying on the upper branch. Moreover the eigenvalue does not vanish at this point, but is imaginary. Whether this corresponds to a Hopf bifurcation or to extinction will be discussed in [14].

FLAME BALLS WHEN Le = 1

Flame balls have only been observed when Le < 0.5, and it is natural to ask whether there are theoretical reasons for this. Here we shall examine the one-dimensional stability when Le = 1.

The equations are (with $R_{ref} = R_0$)

$$\frac{\partial \overline{T}}{\partial \overline{t}} = \frac{1}{\overline{r}^2} \frac{\partial}{\partial \overline{r}} (\overline{r}^2 \frac{\partial \overline{T}}{\partial \overline{r}}) + \overline{B} \, e^{-\theta/2\overline{T}_*} \, \delta(\overline{r} - \overline{r}_*) - \frac{1}{\theta} F(\overline{T}),$$

$$\frac{\partial Y}{\partial \bar{t}} = \frac{1}{\bar{r}^2}\frac{\partial}{\partial \bar{r}}\left(\bar{r}^2 \frac{\partial Y}{\partial \bar{r}}\right) - \bar{B}\, e^{-\theta/2\bar{T}_*}\, \delta(\bar{r} - \bar{r}_*), \tag{26}$$

where $F \equiv 0$ for $\bar{r} > \bar{r}^*$.

The steady solution is:

$\bar{r} < \bar{r}^*$ (=1)

$$Y = 0, \quad \bar{T} = \bar{T}_a + \frac{1}{\theta}\left(\frac{1}{6}\bar{r}^2 F_a - \frac{1}{2} F_a\right)$$

$\bar{r} > 1$

$$Y = Y_\infty\left(1 - \frac{1}{\bar{r}}\right), \quad \bar{T} = \bar{T}_\infty + \frac{1}{\bar{r}}\left(Y_\infty - \frac{1}{3\theta} F_a\right) \tag{27}$$

where $F_a \equiv F(\bar{T}_a)$.

$$\bar{T}_* = \bar{T}_a - \frac{1}{\theta} \cdot \frac{F_a}{3} \tag{28}$$

is the flame-temperature, and the steady radius R_0 is defined by

$$Y_\infty = \bar{B}\, e^{-\theta/2\bar{T}_a} \cdot e^{-F_a/6\bar{T}_a^2} \tag{29}$$

which is equivalent to (23).

Infinitesimal perturbations have the form:

$\bar{r} > \bar{r}^*$

$$T' = \frac{C_1}{\bar{r}}\, e^{-\sqrt{\alpha}\bar{r} + \alpha\bar{t}}, \quad Y' = \frac{C_2}{\bar{r}}\, e^{-\sqrt{\alpha}\bar{r} + \alpha\bar{t}}$$

$\bar{r} < \bar{r}^*$

$$T' = \frac{B_1}{\theta}\, \frac{\sinh(\sqrt{\alpha}\,\bar{r})}{\bar{r}} \cdot e^{\alpha\bar{t}}, \quad Y' = 0 \tag{30}$$

$$|\arg\alpha| < \pi$$

where we have anticipated the relative magnitude of the temperature perturbations within the flame-ball and so, because $0(1/\theta)$ terms can be neglected, have ignored the contribution that F makes to T'.

If the perturbed flame-sheet is

$$\bar{r}_* = 1 + \delta\, e^{\alpha \bar{t}}\,, \tag{31}$$

continuity of $\bar{T}$ and Y leads to the equations:

$$\begin{aligned} &C_1\, e^{-\sqrt{\alpha}} - \frac{B_1}{\theta} \sinh \sqrt{\alpha} - \delta Y_\infty = 0, \\ &\delta Y_\infty + C_2\, e^{-\sqrt{\alpha}} = 0. \end{aligned} \tag{32}$$

The perturbation flame-temperature is

$$T_*' = -\delta\, e^{\alpha \bar{t}} \left(Y_\infty - \frac{F_a}{3\theta}\right) + C_1\, e^{-\sqrt{\alpha} + \alpha \bar{t}} \tag{33}$$

so that the gradient conditions yield:

$$\begin{aligned} &(C_1 + C_2)(1 + \sqrt{\alpha})\; e^{-\sqrt{\alpha}} + \frac{B_1}{\theta}(\sqrt{\alpha} \cosh \sqrt{\alpha} - \sinh \sqrt{\alpha}) + \frac{\delta F_a}{\theta} = 0, \\ &\frac{2T_*^2}{Y_\infty \theta}\left(-2\delta Y_\infty - C_2\,(1 + \sqrt{\alpha})\, e^{-\sqrt{\alpha}}\right) = -\,\delta\left(Y_\infty - \frac{F_a}{3\theta}\right) + C_1\, e^{-\sqrt{\alpha}}\,. \end{aligned} \tag{34}$$

The homogeneous system (33), (34) only has a solution if α satisfies a dispersion relation, and in the limit $\theta \to \infty$ this is

$$\begin{aligned} &D(\sqrt{\alpha}, \beta) \equiv \sqrt{\alpha}\; e^{\sqrt{\alpha}}\, (\beta + 1 - \sqrt{\alpha}) - 3\beta \sinh \sqrt{\alpha} = 0\,, \\ &\beta \equiv \ln\,(R_0/R_z)\,. \end{aligned} \tag{35}$$

β changes monotonically between 0 and ∞ as the steady solution curve is traversed.

When $\beta = 0$ there is an unstable root and it is easy to follow its behavior as β is increased (Fig. 3). The transition from real to complex values occurs when $\frac{\partial D}{\partial\sqrt{\alpha}} = 0$, i.e.,

$$e^{\sqrt{\alpha}}(\alpha - 3\sqrt{\alpha} + 3/2) + e^{-\sqrt{\alpha}}(3\alpha - 3/2) = 0 \tag{36}$$

with solutions $\sqrt{\alpha} = .408..$ ($\beta = .563..$, point B of Fig. 3), $\sqrt{\alpha} = 2.276$ ($\beta = 3.668..$, point D of Fig. 3). As $\beta \to \infty$ we have the estimate $\sqrt{\alpha} \sim \beta - 1/2 + \cdots$. Since $|\arg\sqrt{\alpha}| < \frac{\pi}{4}$ for all β, all stationary solutions are unstable to one-dimensional disturbances. This result, combined with the results of [10] described in the previous section means that there is a critical Lewis number $Le_c < 1$, and for $Le < Le_c$ there are stable solutions, for $Le_c < Le \leq 1$ there are none.

ACKNOWLEDGMENT

This work was partly supported by the Air Force Office of Scientific Research, the National Science Foundation, the NASA Lewis Research Center, and the Gas Research Institute.

REFERENCES

[1] Ronney, P. D. "Near-Limit Flame Structures at Low Lewis Number," to appear in Combustion and Flame (1990).

[2] Ronney, P. D., Combustion Science and Technology, 59, 123 (1988).

[3] Ronney, P. D. and Wachman, H. Y., Combustion and Flame, 62, 107 (1985).

[4] Ronney, P. D., Combustion and Flame, 62, 121 (1985).

[5] Ronney, P. D., "On the Mechanisms of Flame Propagation Limits and Extinguishment Processes at Microgravity," 22nd Symposium (International) on Combustion, The Combustion Institute, 1615 (1988).

[6] Ronney, P. D., Acta Astronautica, 12, No. 11, 915 (1985).

[7] Ronney, P. D., "A Study of the Propagation, Dynamics, and Extinguishments of Cellular Flames Using Microgravity Techniques," AIAA Paper No. 89-0157, 27th Aerospace Sciences Meeting, January 1989.

[8] Ronney, P. D. and Sivashinsky, G. I., SIAM J. on Applied Mathematics, 49, 1029 (1989).

[9] Buckmaster, J. and Joulin, G., "Radial Propagation of Premixed Flames and $\sqrt{t}$ behavior," Combustion and Flame, 78, 275 (1989).

[10] Buckmaster, J., Joulin, G. and Ronney, P. D., "The Structure and Stability of Non-Adiabatic Flame-Balls," Combustion and Flame, 79, 381 (1990).

[11] Buckmaster, J. D. and Ludford, G. S. S., Theory of Laminar Flames, U. of Cambridge Press, p.11 , (1982).

[12] Lewis, B. and von Elbe, G., Combustion, Flames, and Explosions of Gases, Academic Press, NY, p. 314 (1961).

[13] see [11], p.47.

[14] Paper in preparation.

NOMENCLATURE

B	pre-exponential factor
C_p	specific heat
E	activation energy
f	volumetric heat loss
F	non-dimensional f
Le	Lewis number
q	integrated volumetric heat loss
Q	heat generated by reaction
r	radius
R	gas constant
R_0	steady flame radius
R_{ref}	reference radius
R_z	Zeldovich flame radius
R_*	$\overline{r_*}/\theta$
t	time
T	temperature
T_a	adiabatic flame temperature $T_\infty + QY_\infty/C_p$
T_b	$T_\infty + QY_\infty/C_p\, Le$
x	$\bar{r} - \overline{r_*}$
Y	mass fraction
α	perturbations $\sim e^{\alpha\bar{t}}$
β	$\ln(R_0/R_z)$
η	$\bar{r}/\bar{t}^{1/2}$
θ	EC_p/RQ
λ	heat conduction coefficient
ρ	density
σ	$\theta^{-1} \ln \bar{t}$
t	$\bar{t}/\theta^2$ or $\bar{t}/\theta$
$(\,)_*$	flame-sheet
$(\,)_\infty$	infinity
$\overline{(\,)}$	non-dimensional

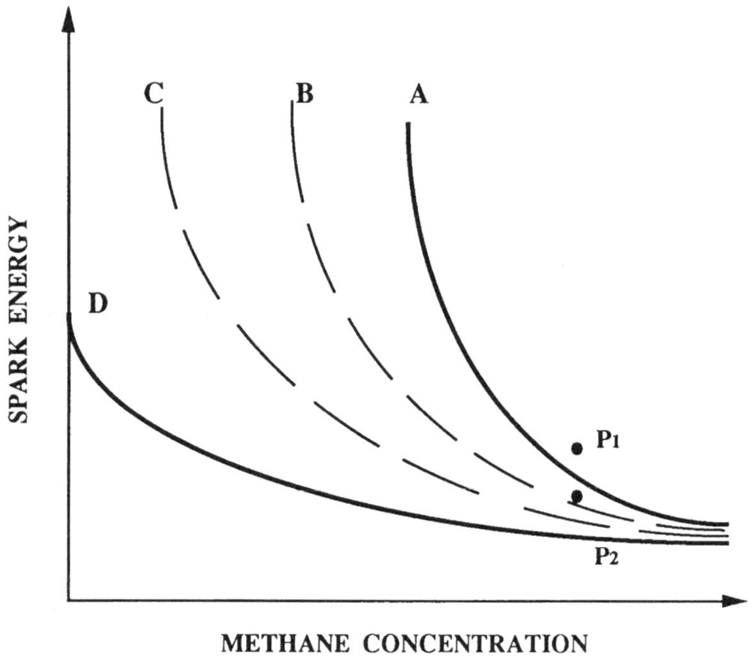

Fig 1. Schematic of Spark Energy - Methane Concentration Curve

A : 0g Minimum Ignition Energy

B,C : Fixed Extinction - Radius Curves for Self - Extinguishing Flames (SEF's)

D : 1g Minimum Igniition Energy

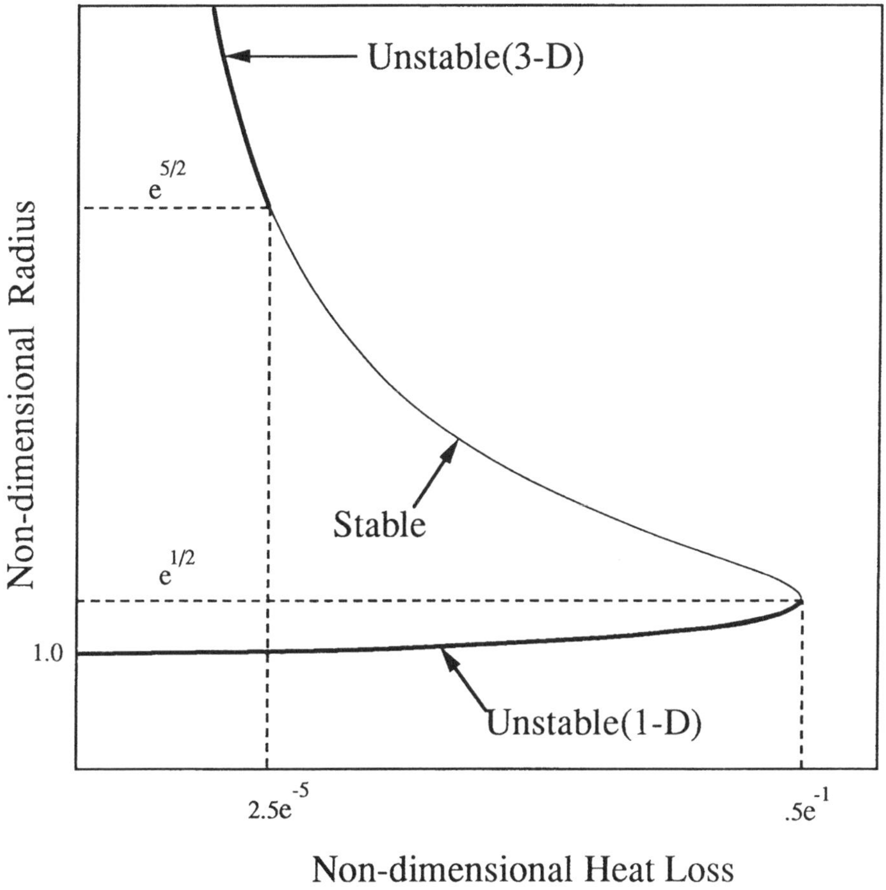

Fig.2. Stationary flame-ball radius R_0/R_z as a function of heat loss $q\,R_z^2$ showing regimes of stability.

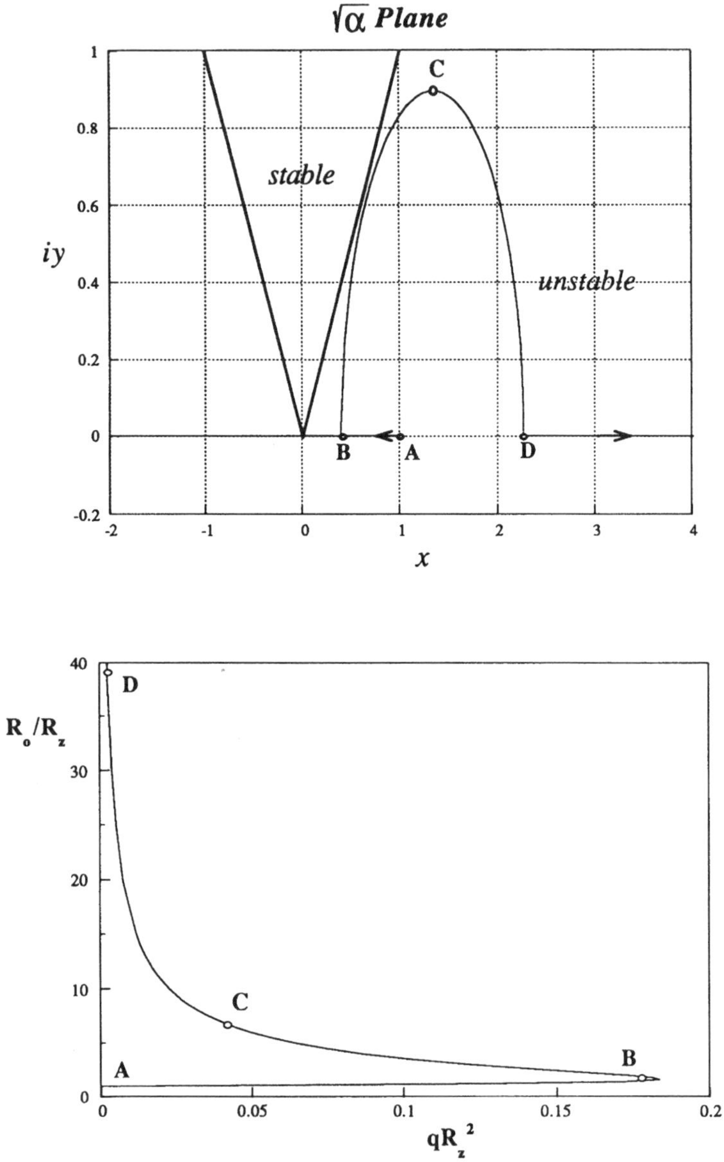

Fig. 3 Locus of $\sqrt{\alpha}$: the motion of $\sqrt{\alpha}$ (=x+iy) in the complex plane is shown as the steady solution curve is traversed.

COHERENT FLAMELET MODEL: APPLICATIONS AND RECENT EXTENSIONS

Sébastien Candel, Denis Veynante, François Lacas, Eric Maistret, Nasser Darabiha and Thierry Poinsot*

E.M2.C. Laboratory, CNRS, Ecole Centrale Paris,
92295 Chatenay-Malabry Cedex, FRANCE
*and SNECMA-Centre de Villaroche-77550 Moissy Cramayel

ABSTRACT. Turbulent combustion poses many difficult problems that are still not completely solved. Physical models of turbulent flames may be constructed in many different ways. The classical approach is based on statistical techniques and the determination of the mean reaction rates usually involves a probability density function which is presumed or determined from as the solution of an evolution equation. Alternative descriptions of turbulent combustion rely on the flamelet concept. The reactive flow field is viewed as a collection of flame elements. The structure of these flamelets is identified and analysed separately. An important advantage of this concept is that it decouples complex chemistry and molecular transport problems from the modeling of the turbulent flow field. A variety of flamelet models has been devised in recent years. In certain cases a probability density function is used to couple the local flamelet analysis to the flow description. Other models use the flamelet passage frequency to evaluate the mean reaction terms. Still another approach is based on a balance equation for the flame area. This equation describes the transport of the flame surface by the turbulent flow field and the physical mechanisms which produce and destroy the reactive surface. This idea originally proposed by Marble and Broadwell has been the subject of numerous developments and extensions in our laboratory. The coherent flamelet description which has evolved from these studies is described in this paper. We first consider some of the fundamental mechanisms which govern turbulent combustion. The balance equation for the flame surface density is first derived along two different paths. We then focus on the premixed version of the coherent flamelet model Applications to a flame stabilized in a duct are then described. A recent extension of the model to situations involving premixed and non-premixed flame sheets. This case which has technological relevance is handled with a coupled set of balance equations for three surface densities. The basic elements of this combined flamelet model are introduced and an illustrative calculation is provided in the case of a reactive shear layer.

CONTENTS

1. Introduction
2. Fundamental processes of turbulent combustion
3. Coherent flamelet model for premixed combustion
4. Experiments on premixed turbulent ducted flames
5. Numerical results and comparison with experiments
6. Configurations combining premixed and nonpremixed flames

7. Combined flamelet model
8. Application to a reactive shear layer
9. Conclusion

1. INTRODUCTION

The problem of turbulent combustion has attracted considerable attention in recent years and some progress has been accomplished in this field. A variety of models is now available and successful calculations have been conducted in simple and more complex configurations. This subject is reviewed in many books and papers (see for example Libby and Williams 1980 and Williams 1985 for basic presentations and Correa and Shyy 1987 for a recent survey of computational models of continuous turbulent combustion). Standard models have mostly relied on statistical techniques and stochastic concepts. The determination of the local average reaction rates has usually involved a probability density function. This pdf is often specified empirically or it may be obtained as a solution of an evolution equation. The latter approach is conceptually attractive because it is based on a framework of stochastic principles but it is not very practical especially if complex chemistry is to be considered. Furthermore the calculation of the pdf is never carried with an exact evolution equation and a set of closure assumptions is required. One drawback of pdf models is that they do not distinguish the fluctuations existing in a turbulent flow in terms of their scale. Large scale, intermediate scale or small scale eddies are thus treated indifferently while their effects on the flame are certainly not comparable. This important issue is not broadly understood.

Alternative descriptions of turbulent combustion have been founded on the flamelet concept. This term designates a variety of physical models in which the reactive flow field is viewed as a collection of flame elements embedded in the turbulent stream (Carrier et al. 1975, Williams 1975, Marble and Broadwell 1977, 1979, Clavin and Williams 1982, Peters 1984, 1986, Spalding 1978, Bray et al. 1984, Bray 1987). Fig. 1 provides a schematic representation of the flamelet ideas. In essence, the instantaneous flame is described as a set of localized reactive sheets. In some restricted views of the flamelet concept the flame sheet should remain uninterrupted and the local instantaneous structure of the flow should be precisely that of a laminar flame. An extended view of the concept allows for discontinuities in the flame sheet and it is then assumed that the reactive sheets are distributed in the flowfield and that their structure may be identified and analysed separately. An advantage of the flamelet concept is that it decouples complex chemistry and molecular transport problems from the modeling of the turbulent flow field. In practical applications a simplified model for the reactive laminar sheets is devised or a flamelet library is first constructed providing specific properties such as the consumption rates per unit flame area, ignition and extinction conditions that are required in the computation of the turbulent flow field.

While the domain of application of flamelet models is not well defined at this time and remains a subject of discussion it is believed that the concept is applicable in the range of large Damköhler numbers and for characteristic turbulent scales much larger than a typical flame thickness. The two conditions are satisfied in many practical situations and flamelet descriptions are probably relevant in many cases and at least in certain ranges of operation of IC engines and continuous flow combustors. The limits of the flamelet regime have been defined up to now on purely intuitive grounds (Barrère 1974, Borghi 1985, Williams 1985, Peters 1986) but new results are available from direct numerical simulations and a revision of current criteria and beliefs is required (see Poinsot et al. 1990 a and b for an analysis of quenching and combustion regimes in turbulent flows). This question is still open but recent evidence indicates that the flamelet concept may be used over a domain which is probably wider than previously thought. Under these circumstances it is reasonable to leave this discussion aside and evaluate the flamelet models from their predictions in specific situations.

Now, there are diverse approaches to the flamelet description. However they all share the following ingredients :

(1) A laminar flamelet submodel (or submodels) providing the local structure and properties of the reactive elements.

(2) A description of the turbulent flow comprising mass average equations describing the mean flow variables and the main species mass fractions and relevant closure equations.

(3) A rule or a set of rules which couple the flamelet submodels to the turbulent flow description.

(4) Additional submodels accounting for chemical reactions taking place outside the flamelets.

There is considerable flexibility in specifying and combining these elements and many alternative ways are being explored. Reviews due to Peters (1984, 1986) and Bray (1987) describe some of the possibilities but they are incomplete and do not make reference to models based on the flame surface density concept which will be emphasized in the present article.

In certain flamelet models a probability density function is used to couple the local analysis to the flow description. This idea is explored by Liew et al (1984) and Peters (1984) to deal with turbulent diffusion flames. It is first implied that at a given point in space the instantaneous structure of the turbulent flame is that of a laminar reactive sheet. Each individual flame element belongs to a family and may be characterized by the reactant composition and the strain rate that is experienced by the flame element. Accordingly it is argued that tere is a unique definition between a scalar quantity ϕ and the normalised conserved scalar ξ : $\phi=\phi(\xi,\chi_{st})$. In this expression χ_{st} designates the scalar dissipation at the stoichiometric point. This quantity has dimensions of s^{-1} and it may be related to the strain rate. The couple (ξ,χ_{st}) may be treated as a random variable having a joined pdf and this function must be presumed or calculated. One usually assumes that the conserved scalar and the scalar dssipation are independent stochastic variables so that the joined pdf may be written as a product: $P(\xi,\chi_{st})=P_1(\xi)P_2(\chi_{st})$. The shape of the conserved scalar pdf P_1 is presumed and its coefficients are determined from the moments of the conserved scalar. The scalar dssipation pdf P_2 is taken as a log-normal distribution and its coefficients are determined from the moments of χ which are themselves related to the other moments of the turbulent variables.

Other models use the flamelet passage frequency to evaluate the mean reaction terms (see Bray 1987 for a survey of this method). The reaction rate is first written as $\overline{w} = \nu_y w_y$ where ν_y designates the flamelet crossing frequency and w_y is the mean reaction rate per flame crossing. A statistical analysis of the flamelet crossing frequency leads to the following expression for the mean reaction term

$$\overline{w} = \frac{g\bar{c}\,(1-\bar{c}\,)}{\hat{L}_y\,|\,\sigma_y\,|}\,\rho_R S_L^0 I_0 \tag{1}$$

where g is a constant of order 1, c is the Reynolds mean of a reaction progress variable, L_y is an integral scale of the wrinkled flame, σ_y is a geometrical factor, $S_L{}^0$ is a burning velocity of an unstrained flame and I_0 is an integral over the strain rate

$$I_0 = \int_0^\infty \frac{W(a)}{W(0)}\,P(a)da \tag{2}$$

where $P(a)$ is the probability density of the strain rate a, $W(a)$ is the consumption rate per unit area of strained flame and $W(0)$ is the consumption rate per unit area of an unstrained flame. The mean consumption rate takes the final form

$$\overline{w} = \frac{g\bar{c}(1-\bar{c})}{\hat{L}_y \mid \sigma_y \mid} \int_0^{\infty} W(a)P(a)da \tag{3}$$

Another class of flamelet models relies on a balance equation for the available flame area. This equation describes the transport of the flame surface by the turbulent flow field and the physical mechanisms which produce and destroy the reactive surface. This approach was originally developed by Marble and his co-workers in an analysis of non-premixed combustion (Carrier, Fendell and Marble 1975, Marble and Broadwell 1977, 1979). It is suggested in these studies that turbulent combustion of unmixed reactants is controlled in the early stages by a competition between straining of the flame elements and mutual annihilation of the flame area due to interactive destruction of neighbouring flame sheets.

The Coherent Flame Model identifies important physical mechanisms of turbulent combustion such as the production of flame area by stretching, its destruction by flame shortening and the central influence of the strain rate acting on the local flamelets (Fig. 2) Because these aspects are important in the large scale coherent motions found in shear flows, the model accounts in some sense for the presence of these organized fluctuations. Our own effort has been to extend the coherent flame description to premixed flow configurations (Candel et al. 1982, Darabiha 1984, Darabiha et al. 1987, Maistret et al. 1989) and to explore its potential in non-premixed situations (Veynante et al. 1987, Lacas et al. 1987). A new description which accounts for premixed and non-premixed flamelets was recently presented (Veynante et al. 1989a). The case of propagating turbulent fronts is treated in separate papers (Lacas et al. 1989, Veynante et al. 1989b). Non-uniformly premixed flames are considered by Veynante et al. (1989c). Some of this work is reviewed in the present paper. Comparisons between calculations and experiments have now been conducted systematically and a few of the results obtained are discussed in this article. The turbulent ducted flame configuration is specifically considered . This case constitutes one of the simplest geometry that may be studied and that has both fundamental and practical interest.

Basic investigations of this flame configuration were conducted by Zukoski and Marble (1955), Wright (1959) and provided considerable information on the complex process of flame stabilization. A qualitative understanding of the problem was also derived from a wide set of careful experiments. Simple scaling laws were obtained which effectively described blowoff conditions. These results provided basic tools in the design of jet engine afterburners. The turbulent ducted flame stabilized on a bluff body still remains a fundamental configuration and its modeling is a subject of continued interest and technological significance. Much of the current work in this field has been directed at the numerical modeling of turbulent flames (see Correa and Shyy 1987 for a recent survey). Modern optical diagnostics and digital signal processing methods have also allowed detailed experiments which could not be carried in the early fifties.

Our own work described in this paper and in other recent articles complement the information on turbulent ducted flames and develop numerical descriptions of this basic configuration. While most of the recent experiments (see Libby et al. 1986 for a review of premixed turbulent flame experiments and Heitor et al. 1988 for a study of baffle stabilized flames) focus on measurements of temperature, velocity or concentration profiles, our effort has been to gather information on the mean source terms and especially on the mean heat release distribution in the flow. There are good reasons which justify this choice. First, it is found that the spatial distributions of the mean heat release are influenced by the equivalence ratio and the flow velocity thus providing a unique view of the interaction between the chemical kinetics and the flow turbulence. Second the data also allow a direct evaluation of the theoretical expressions used to model the mean consumption rates appearing in the averaged balance equations. Of course we are also perfectly aware that standard measurements of the flow variables and of their fluctuations are also valuable but we wish to show that much can be learnt from a less standard type of analysis.

The detailed examination of the mean source term differs from the more common tests performed in the literature on velocity, temperature and mass fraction profiles all of which are only indirectly related to the modeled reaction rates. Because the flow variables are obtained by integrating the dynamic equations they are less sensitive to the modeling assumptions and do not allow a direct assessment of the combustion models. It is a fact that reasonable mean flow profiles may be obtained with the simplest assumptions.

It is then difficult to see if the model is adequate . If one wishes to describe the effects of finite rate chemistry and turbulence on the structure of the flame it appears clearly that the mean source terms should be examined and precisely represented. With this goal in mind a large set of experiments has been carried to measure the distribution of light emission from C_2 and CH radicals. The spatial distributions of the radiated light may be interpreted as giving a qualitative mapping of the local mean heat release in the turbulent flame. Because finite-rate chemistry effects are quite pronounced the modeling of the flow is a challenging problem. Indeed standard "fast" chemistry models are unable to reproduce the trends observed. Improved descriptions are needed which account explicitly for the interaction between the chemical kinetics and the flow.

We begin with a rapid survey of some of the processes which are believed to be important in turbulent combustion (Section 2). The special case of premixed combustion is considered in Section 3. Experimental results are described in Section 4 and compared with numerical calculations in Section 5. Turbulent flames containing premixed and nonpremixed reactive elements are described in Section 6. The physical processes which govern these situations are discussed and modeled in Section 7. An illustration of the combined flamelet model is provided in Section 8.

2. FUNDAMENTAL PROCESSES OF TURBULENT COMBUSTION

A large variety of processes govern turbulent combustion. Many of these processes are not fully understood at this time but it appears possible to describe some of the mechanisms on an essentially intuitive basis complemented by selected experimental data, analytical studies and numerical simulations. We will only consider flamelet regimes of combustion and adopt the view that the reactive stream contains a collection of laminar flame sheets. This situation is typified by the spark Schlieren photograph displayed in Figure 3. This picture shows the instantaneous structure of a ducted flame stabilized on a bluff body. Large scale vortices are made apparent by the large temperature gradients existing between the cold fresh mixture and hot combustion products flowing behind the stabilizer. These large structures are observed in a large range of Reynolds numbers (see Darabiha et al.1987 for Schlieren photographs corresponding to different flow regimes). The vortices observed in reactive shear flows have much in common with those found in the classical experiment of Brown and Roshko (1974) but there are some distinguishing features. It is found in particular that the initial vortices formed near the stabilizer lip are perfectly rolled up in the form of "cat's eye" patterns while at greater dstances from the origin of the shear layer the inner boundary of the vortex patterns disappears and a core of burnt products is formed on the hot side of the flame. This evolution is directly related to the competition which takes place inside the flow between mechanisms which tend to augment the amount of flame surface and mechanisms which tend to destroy or dissipate the flame area. The increase of flame area is essentially due to the strain rates acting in the plane of the flame while its dissipation is dominated by the consumption of the fresh mixture and the mutual annihilation of adjacent reactive sheets. In addition one expects that excessively high rates of strain will prevent the extension of the flame and could even lead to localized extinctions. Consequently some of the turbulent strain rates may be responsible for a reduction of the flame surface. The conditions which must prevail for this last mechanism are not fully known at this time. It is argued by some authors that extinction begins at the smallest scales (the Kolmogorov scales) existing in the flow but the direct simulations of Poinsot et al. (1990) seem to indicate that this process takes place in a range of intermediate and more energetic scales.

A dynamic representation which fits to the configuration found in Figure and leads to a detailed balance of the flame elements is provided by the Coherent Flame Model by the evolution and are conveniently described in the simple terms of the Coherent Flame Model. It is assumed that the flame elements are convected and distorted by the turbulent motion but that they retain an identifiable structure. In this sense, the flamelets remain "coherent" (i.e. "organized"). This is the case if the thickness of the reaction zone δ_r is sufficiently small with respect to the typical scales l_t of the energy containing turbulent motion. It is believed by some authors that δ_r should be greater than all the scales existing in the turbulent flow i.e. that it should be greater than the Kolmogorov scale. However recent numerical simulations (Poinsot et al. 1990) indicate that there exists a cut-off scale in each turbulent flow. Fluctuations which have a size smaller than this scale are unable to interact with the flame. As a consequence the flame thickness could certainly exceed the Kolmogorov scale and the flow would still be in the flamelet regime.

A rough sketch of the flow configuration is given in the case of a reactive shear layer (Figure 2). A flame element placed in the flowfield is mainly affected by the local strain rate which acts in the plane of the flame, modifies its structure and changes the local reaction rate. As a consequence, the local consumption rate per unit of flame area may be obtained from an analysis of strained laminar premixed or diffusion flames. The simplest geometry allowing this analysis is shown in Figure 2-c. The strain rate imposed by the turbulent motion has also the effect of increasing the available flame area. This production process is balanced by various destruction mechanisms such as flame shortening (mutual annihilation of adjacent flamelets by consumption of one of the reactant, Figure 2-b) or flame quenching (due, for example, to an excessively large strain rate). This mechanism is well illustrated in by a generic problem devised by Marble (1985) and Karagozian and Marble (1986). One considers an initially plane flame interacting with a vortex. The flame is wound up by the flow field and flame area is created by the local strain rates acting on the flame sheet. The local production of surface is actually proportional to the tangent strain component. The flame elements formed near the vortex core interact and the intervening reactant is rapidly consumed giving place to a burnt core of products and in this region the flame surface disappears. The process is illustrated by calculations performed by Laverdant and Candel (1988 and 1989). Typical results are shown in Fig. 4 which corresponds to non-premixed and premixed flame situations.

It follows from these general considerations that the mean comsumption rate of a species i per unit volume at a point of the flow, W_i, may be determined as the product of the mean flame surface density at that point (i.e. the flame surface per unit of volume), Σ_f, by the consumption rate per unit of flame area, V_{Di}, obtained from the analysis of local strained laminar flamelets :

$$W_i = \rho V_{Di}\, \Sigma_f$$

In its simplest form, the coherent flame description of turbulent reactive flows combines the three following elements : (1) a model for the turbulent flow comprising a standard set of Reynolds or mass average dynamic equations and a turbulence closure model (2) a local model for the laminar strained flame elements providing the consumption rates per unit flame area. Complex molecular transport and detailed chemistry may be incorporated into this model which is essentially decoupled from the turbulent flow calculation (3) a balance equation for the flame surface density accounting for transport, diffusion, production and destruction of flame area.

The specificity of the Coherent Flame Model lies in the use of the *flame surface density* to relate the local flamelet analysis to the global turbulent flow field calculation. This equation takes into account transport, diffusion, production and destruction of flame area and has the general form :

{ transport } = *{ turbulent diffusion }* + *{ production }* -*{destruction}*

A balance equation of this type is proposed by Marble and Broadwell to describe the evolution of the flame surface density in the case of a turbulent diffusion flame. While the balance equation for the flame surface was initially devised on the basis of some intuitive arguments it may be derived from basic principles (see below). Considering a non-premixed flame involving a global reaction between the two main species F and O :

$$F + s\,O \text{ ----> } P$$

the balance of flame surface may be written :

$$\frac{\partial \Sigma_f}{\partial t} + v_k \frac{\partial \Sigma_f}{\partial x_k} = \frac{\partial}{\partial x_k}\left[D \frac{\partial \Sigma_f}{\partial x_k}\right] + \varepsilon_s \Sigma_f - \left(\frac{V_{D_F}}{X_F} + \frac{V_{D_O}}{X_O}\right) \Sigma_f^2 \tag{4}$$

whereΣ designates the flame surface density, D is a turbulent diffusion coefficient, V_{DO} and V_{DF} are the volume rates of consumption of oxydizer and fuel per unit flame area and X_O and X_F are the oxidizer and fuel mole fractions.

The production term describes the increase of flame area due to the local strain rate ε_s whereas the destruction term only takes into account the flame shortening mechanism due to mutual interaction of adjacent flame elements. The local strain rate plays an important role in this flamelet model and must be derived from known variables characterizing the turbulent flow (turbulent kinetic energy, dissipation, typical turbulent length scale, ...). In the case of a two dimensional mixing layer such as that shown in Figure an estimate of the local strain rate ε_s may be deduced fromed the transverse gradient of the mean axial velocity : $\varepsilon_S = \alpha\,|\partial u / \partial y|$.

This expression is applicable in simple flows like shear layers but it is not adequate in more complex situations.

3. COHERENT FLAMELET MODEL FOR PREMIXED COMBUSTION

3.1 Balance Equation for the Flame Area per Unit Mass

The balance equation for the flame area plays a key role in this model. This equation describes the competition between the various processes which create or destroy flame area. It may be derived from first principles but some additional closure assumptions are also required. In the basic Coherent Flame Model the balance equation relating the local and global description levels was written for the flame area per unit volume designated asΣ_f. Now, our recent applications to reactive flows with nonsteady changes of the mean pressure as found for example in studies of internal combustion engines indicate that a slightly different quantity is more adequate. In fact the flame area per unit mass S_f arises quite naturally in the derivation of an instantaneous balance of flame surface. The two quantities are of course related by $\Sigma_f = \rho\, S_f$.

The balance equation for the mean flame area per unit mass may be derived by adopting the same procedure as that used for Σ_f (see Darabiha et al. 1987). It is also possible to follow a different path and start from a general expression for the flame stretch. Consider a flame front (Figure 5) described by the function $F(\boldsymbol{x},t) = 0$. Let

$$\boldsymbol{n} = -\nabla F \,/\, |\nabla F| \tag{5}$$

designate the unit normal to the front. To fix the ideas $F > O$ corresponds to the burnt material while $F < O$ relates to the fresh mixture and consequently the normal n is oriented from the burnt to fresh side. A point belonging to the flame front moves with velocity $\boldsymbol{w}$. This point remains on the surface if

$$\frac{\partial F}{\partial t} + \boldsymbol{w}.\nabla F = 0 \tag{6}$$

If the front propagates in the normal direction at a speed S_L, the velocity $\boldsymbol{w}$ will appear as the sum of the local fluid velocity and the flame speed in the normal direction

$$\boldsymbol{w} = \boldsymbol{v} + S_L \boldsymbol{n} \tag{7}$$

It is shown (see for example Candel and Poinsot 1990 and references therin) that the relative variation of an element of flame may be cast in the form

$$\phi_S = \frac{1}{\delta A}\frac{d(\delta A)}{dt} = -\boldsymbol{nn}:\nabla \boldsymbol{v} + \nabla.\boldsymbol{v} + S_L \nabla.\boldsymbol{n} \tag{8}$$

The first term appearing in this expression is the normal element of the strain rate tensor, the second represents the volumetric expansion of the fluid and the third describes the effects of flame curvature. This last term may be expressed in terms of the principal radii of curvature of the surface

$$S_L \nabla . \boldsymbol{n} = S_L(\frac{1}{R_1} + \frac{1}{R_2}) \tag{9}$$

It is convenient and more meaningful to use a slightly different form for the first two terms of the stretch. For this let us introduce the components ν_j and η_j are components of two orthogonal unit vectors ν and η belonging to the local tangent plane. The unit vector in the normal direction is then given by

$$\boldsymbol{n} = \nu \times \eta$$

The following identity is easily established:

$$\nu_j \nu_l + \eta_j \eta_l = \delta_{jl} - n_j n_l \tag{10}$$

Substituting this relation in expression (8) one finds the alternative form

$$\phi_S = \frac{1}{\delta A} \frac{d(\delta A)}{dt} = (\nu\nu + \eta\eta) : \nabla \boldsymbol{v} + S_L \nabla . \boldsymbol{n} \tag{11}$$

The flame is stretched by the rates of strain associated with the flow and acting in the tangent plane of the flame and by the normal propagation of the front. A plane flame propagating in a uniform flow does not experience stretch. A plane flame established in the vicinity of a stagnation point is stretched by the strain rate associated with this type of flow. Other configurations producing flame stretch are discussed by Matalon (1983), Strehlow and Savage (1978), Buckmaster (1979, 1982).

Let us now consider the flame surface density $\Sigma = \delta A / \delta V$. To calculate the rate of change of Σ as the front moves in the fluid we need an expression for the rate of change of a volume element moving with the flame at the velocity $\boldsymbol{w}$. If the front propagates in the normal direction at a speed S_L, this velocity $\boldsymbol{w}$ is the sum of the local flow velocity and the flame speed in the normal direction

$$\boldsymbol{w} = \boldsymbol{v} + S_L \boldsymbol{n} \tag{12}$$

The fractional rate of change of δV is easily deduced from the transport theorem for a moving volume $V(t)$

$$\frac{d}{dt}\int_{V(t)} f dV = \int_{V(t)} \frac{\partial f}{\partial t} dV + \int_{S(t)} f \boldsymbol{w}.\boldsymbol{n} dA \tag{13}$$

Substituting $f = 1$ in this expression yields

$$\frac{d}{dt}\int_{V(t)} dV = \int_{S(t)} \boldsymbol{w}.\boldsymbol{n} dA \tag{14}$$

or equivalently

$$\frac{d}{dt}\int_{V(t)} dV = \int_{V(t)} \nabla.\boldsymbol{w} dV \tag{15}$$

For a volume element δV this expression becomes

$$\frac{1}{\delta V}\frac{d(\delta V)}{dt} = \nabla.\boldsymbol{w} \tag{16}$$

Combining this result with expression (8) one finds

$$\frac{1}{\Sigma}\frac{d\Sigma}{dt} = -\boldsymbol{nn}:\nabla \boldsymbol{v} - \boldsymbol{n}.\nabla S_L \tag{17}$$

It is important to note that the operator d/dt appearing in these expressions involves a convective derivative based on the total flame speed $\boldsymbol{w} = \boldsymbol{v} + S_L \boldsymbol{n}$:

$$\frac{d}{dt} = \frac{\partial}{\partial t} + \boldsymbol{w}.\nabla$$

Expression (17) may be written in the equivalent form

$$\frac{\partial \Sigma}{\partial t} + \nabla.\boldsymbol{v}\ \Sigma = —(\boldsymbol{nn}:\nabla \boldsymbol{v} — \nabla.\boldsymbol{v})\Sigma — \boldsymbol{n}.\nabla(S_L \Sigma) \tag{18}$$

One may also replace $—(\boldsymbol{nn}:\nabla v — \nabla.\boldsymbol{v})$ by $(vv + \eta\eta):\nabla\boldsymbol{v}$ and write

$$\frac{\partial \Sigma}{\partial t} + \nabla.\boldsymbol{v}\ \Sigma = \left[(vv + \eta\eta):\nabla\boldsymbol{v})\right]\Sigma — \boldsymbol{n}.\nabla(S_L \Sigma) \tag{19}$$

While this equation constitutes a local balance for the flame area per unit volume it is often more appropriate to consider the flame area per unit mass. This new density designated by a_f is obtained from Σ by $\Sigma = \rho S_f$. Inserting this definition into (40) yields

$$\frac{\partial(\rho S_f)}{\partial t} + \nabla.\rho\boldsymbol{v}\ S_f = \rho\left[(vv + \eta\eta):\nabla\boldsymbol{v})\right]S_f — \boldsymbol{n}.\nabla(\rho S_L S_f) \tag{20}$$

The effects of flame curvature may be exhibited by writing the previous balance in the form

$$\frac{\partial(\rho S_f)}{\partial t} + \nabla.\rho(\boldsymbol{v} + S_L\boldsymbol{n})S_f = \rho\left[(vv + \eta\eta):\nabla\boldsymbol{v})\right]S_f + \rho(S_L\nabla.\boldsymbol{n})S_f \tag{21}$$

Introducing the strain rate tensor

$$s_{ij} = \frac{1}{2}\left(\frac{\partial v_i}{\partial x_j} + \frac{\partial v_j}{\partial x_i}\right)$$

and using indices the previous expression becomes

$$\frac{\partial}{\partial t}(\rho S_f) + \frac{\partial}{\partial x_k}\left[\rho(v_k + S_L n_k)S_f\right] = \rho\left[(v_i v_j + \eta_i\eta_j)s_{ij}\right]\ S_f + \rho\left(S_L\nabla.\boldsymbol{n}\right)S_f \tag{22}$$

The effect of the strain rate acting in the plane of the flame and of flame curvature now appear explicitely. The combination of these two terms is known as the flame stretch (the fractional variation of flame surface element per unit time). In the following development we will neglect the terms containing the flame speed S_L and we use ε_S to designate the strain acting in the plane of the flame

$$\varepsilon_S = (\nu_i \nu_j + \eta_i \eta_j) s_{ij}$$

These assumptions are adopted for simplicity but it is possible to write a more exact average equation including the flame relative motion and the curvature terms. Introducing the mass average decomposition of the different variables appearing in the balance equation for the flame surface one gets

$$\frac{\partial}{\partial t}(\bar{\rho}\tilde{S}) + \frac{\partial}{\partial x_k}(\bar{\rho}\tilde{v}_k\tilde{S}) = \bar{\rho}\,\widetilde{\varepsilon_S S} + \rho\widetilde{\varepsilon_S'' S''} + \frac{\partial}{\partial x_k}\left(-\bar{\rho}\,\widetilde{v_k'' S''}\right) \tag{23}$$

At this point it is necessary to make use of closure assumptions. A standard expression for the turbulent flux term is:

$$-\bar{\rho}\,\widetilde{v_k'' S''} = \frac{\mu_t}{\sigma_s}\frac{\partial \tilde{S}}{\partial x_k}$$

Now the first two terms may be estimated as $\bar{\rho}\,\widetilde{\varepsilon_S S} + \rho\widetilde{\varepsilon_S'' S''} = \alpha\bar{\rho}\tilde{\varepsilon}_S\tilde{S}$ where ε_S is a mean strain rate. This term describes the *augmentation of material surface by the rate of strain*. This expression of the production term differs slightly from that used in our previous work because we now consider the flame surface per unit mass.

These assumptions are sufficient to close the balance of mean flame surface per unit mass S. However certain processes are not described by the previous equation. More specifically it is necessary to include a *chemical shortening term* similar to that appearing in the balance equation for the flame surface density Σ_f. This destruction term describing the process of flame shortening (the destruction of flame area by consumption of the reactants separating adjacent flame sheets) may take different forms. For premixed flames, we consider that the fractional rate of annihilation of flame surface is proportional to the mass of fuel burnt per second and inversely proportional to the mass of fuel in the control volume V considered :

$$\left(\frac{1}{\overline{\rho}\widetilde{S}_f}\frac{d\overline{\rho}\widetilde{S}_f}{dt}\right)_{annihilation} = -\beta\frac{\overline{\rho}Q\widetilde{S}_f V/(-\Delta h_f^0)}{\overline{\rho}\widetilde{Y}_F V} \tag{24}$$

where the heat release rate per unit flame area Q is obtained from the local flamelet calculations and $\left(-\Delta h_F^o\right)$ designates the heat release per unit mass of fuel. The annihilation term is then written as :

$$-\beta\,\overline{\rho}\,\frac{Q}{(-\Delta h_f^0)\,\widetilde{Y}_F}\,\widetilde{S}_f^2 \tag{25}$$

and the balance equation of flame surface per unit mass becomes

$$\frac{\partial}{\partial t}\left(\overline{\rho}\widetilde{S}_f\right) + \frac{\partial}{\partial x_k}\left(\overline{\rho}\widetilde{v}_k\widetilde{S}_f\right) = \frac{\partial}{\partial x_k}\left[\frac{\mu_t}{\sigma_{S_f}}\frac{\partial \widetilde{S}_f}{\partial x_k}\right] + \alpha\overline{\rho}\,\widetilde{\varepsilon}_s\,\widetilde{S}_f - \beta\overline{\rho}\,\frac{Q}{\left(-\Delta h_F^o\right)\widetilde{Y}_F}\,\widetilde{S}_f^2$$

Numerical simulations based on this equation as well as physical arguments also indicate that the production term describing the effect of the strain rate should be modified to take into account limiting mechanisms like *extinction*. Furthermore the determination of the effective strain rate should take into account the scale of the turbulent eddies which are acting on the flame surface. It is shown by Poinsot et al.(1990 a and b) that certain turbulent eddies are too small to interact efficiently with the flame and hence do not participate to the production of flame area by strain. There is a cut-off scale below which the turbulent fluctuations do not have a noticeable effect on the flame. It is also shown that quenching of the flame may take place in a range intermediate turbulent scales. When this process takes place some of the turbulent eddies (essentially the large scale eddies) participate to the production of flame area by the strain they impose to the reactive sheet while other turbulent scales have a contrary effect and tend to diminish the flame surface by quenching some portions of the reaction sheets. The production of flame surface associated with the strain rate is less efficient as this quantity strain rate takes large values. In a field of very high strain the process may even become completely ineffective and the strain may then only contribute to the destruction of flame surface. The analytical consequences of the study carried by Poinsot et al. (1990 a) have not yet been fully exploited but it is nevertheless possible to write some simple models to account for these features. One idea is to modify the production term if the strain rate becomes too large. This may be done in various ways. For instance, one may consider a certain critical strain rate ε_{se} (which depends on the local conditions and is determined from the local analysis or from direct simulations of the kind performed by Poinsot et al. 1990a) and write the destruction term as $-2\alpha\rho\left(\varepsilon_s-\varepsilon_{se}\right)h\left(\varepsilon_s-\varepsilon_{se}\right)S_f$ where h is the heavyside function ($h(x)$ $= 1$ if $x>0$ and $h(x) = 0$ if $x\leq 0$).

Another possibility is to multiply the production term by a function representing the efficiency of the strain rate (close to 1 for low strain rates and close to 0 for high strain rates). The fonction Q/Q_0 (where Q is the heat release rate per unit flame area -obtained from the local flamelet analysis- and Q_0 is the value of Q for ε_S=0) is well representative of this influence of the strain rate. The production term becomes $\alpha\rho\,\varepsilon_s S_f \frac{Q}{Q_0}$. This expression is used in the calculations presented in this paper. The balance equation for S_f is finally written as

$$\frac{\partial}{\partial t}\left(\bar{\rho}\widetilde{S_f}\right)+\frac{\partial}{\partial x_k}\left(\bar{\rho}\tilde{v}_k\widetilde{S_f}\right)=\frac{\partial}{\partial x_k}\left[\frac{\mu_t}{\sigma_{S_f}}\frac{\partial\widetilde{S_f}}{\partial x_k}\right]+\alpha\bar{\rho}\,\tilde{\varepsilon}_s\,\widetilde{S_f}\frac{Q}{Q_0}-\beta\bar{\rho}\frac{Q}{\left(-\Delta h_F^o\right)\tilde{Y}_F}\widetilde{S_f}^2 \quad (26)$$

The strain rate appearing in this equation may be evaluated from $\varepsilon_S = C_S\frac{\varepsilon}{k}$ and the constants take the following values : $\sigma_{S_f} = 1$, $\alpha = 10$, $\beta = 0.4$ and $C_S = 0.17$.

It is probable that the final form of the balance equation for the flame area will also differ from that presented in this paper. The consequences of some recent studies like those of Poinsot et al. (1990 a and b) are not yet incorporated in the model and many new ideas are being tested. This is a subject of current research.

At this point it is appropriate to give another derivation of the balance equation for the flame surface. The initial idea is due to Peters et al. (1989) but the calculations given below differ from those contained in that reference. We wish to show that the balance expression obtained previously may also be found by following a distinct line of reasoning .

Consider again the equation describing the propagation of the front (6). This equation may be written as

$$\frac{\partial F}{\partial t} + v.\nabla F = S_L \left| \nabla F \right| \tag{27}$$

This equation is adopted by Kerstein et al. (1988) and Ashurst et al. (1988) to describe the evolution of interfaces in a turbulent flow field. It is shown by these authors that under certain conditions the volume average of $\left| \nabla F \right|$ provides the ratio of the turbulent to the laminar flame speed S_T/S_L. According to Peters et al the ensemble average of $\left| \nabla F \right|$ may be interpreted as the mean flame surface area per unit volume $\bar{\Sigma} = \overline{\left| \nabla F \right|}$. It is then possible to obtain an equation for this quantity from the field equation describing the front propagation. Taking the gradient of this equation one finds:

$$\frac{\partial \nabla F}{\partial t} + \nabla (v.\nabla F) = \nabla (S_L \left| \nabla F \right|) \tag{28}$$

The convective term may be cast in the form

$$\nabla (v.\nabla F) = \nabla F \,.\, \nabla v + v \,.\, \nabla \nabla F + (\nabla \times v) \times \nabla F \tag{29}$$

and equation (28) becomes

$$\frac{\partial \nabla F}{\partial t} + \nabla F \,.\, \nabla v + v \,.\, \nabla \nabla F + (\nabla \times v) \times \nabla F = S_L \nabla \left| \nabla F \right| + \left| \nabla F \right| \nabla S_L \tag{30}$$

Taking the scalar product of this expression by $2\,\nabla F$ one obtains after a few transformations:

$$\frac{\partial (\nabla F)^2}{\partial t} + 2(\boldsymbol{nn} : \nabla v)(\nabla F)^2 + v \,.\, \nabla (\nabla F)^2 = -\, S_L \boldsymbol{n} \,.\nabla (\nabla F)^2 - 2(\nabla F)^2 \boldsymbol{n} \,.\nabla S_L \tag{31}$$

Let us now consider the instantaneous density $\Sigma = \left| \nabla F \right|$. The balance equation for Σ is easily deduced from expression (31):

$$\frac{\partial \Sigma}{\partial t} + (\boldsymbol{v} + S_L \boldsymbol{n}).\nabla S = -(\boldsymbol{nn} : \nabla \boldsymbol{v})\Sigma - \Sigma \boldsymbol{n}.\nabla S_L \tag{32}$$

This equation is exactly identical to that deduced from the stretch expression. This confirms the validity of the equation adopted to describe the transport of the flame surface density. While the instantaneous expression appears now well justified the modeled form providing the mean density is less firmly based. Clearly some of the terms of this equation need further studies and refinements. Other ideas are also worth exploring. One of them consists of using a stochastic comprising an evolution equation for the flame area as proposed by Pope and Cheng (1988). A coupled set of equations describing the effects of the coherent fluctuations and fine grained turbulence on the material surface density is established by Liu (1988) and further work is being carried along this line.

3.2 The Local Flamelet Structure

Consumption rates per unit flame area and the corresponding heat release are obtained from an examination of the local flame structure. Many configurations may be used as local models (Figure 6). One may for example consider the counterflow of fresh reactants and burnt gases as shown in Figure 6-a. An alternative model is that of twin flames formed by the counterflow of two fresh streams as sketched in Figure 6-b. Both models are being considered but only the first is used in the present calculations. Computational methods are described in detail by Giovangigli (1988) and results for propane-air strained flames are contained in Darabiha et al. (1988) for the single flame case and in Djavdan et al (1990) for the symmetric double flame situation.The flamelet structure and properties depend on pressure, fresh gas temperature T_o, equivalence ratio Φ, burnt gas temperature T_b and strain rate ε_s.

A large number of flamelets has been determined for a constant pressure of 1 bar and a constant fresh gas temperature $T_o = 300$ K (see Darabiha et al. 1988). For a given hot stream temperature T_b the heat release per unit flame area Q may be obtained as a function of the equivalence ratio and the strain rate. A typical plot of $Q(\Phi,\varepsilon_s)$ is shown in Figure 7. In this case the temperature T_b is equal to the adiabatic flame temperature $T_b = T_{ad}(\Phi)$. These data are used in the numerical simulations included in this paper. More generally all the flamelet structures and characteristics may be stored in a large data base and the values of the consumption and heat release rates may be retrieved during the calculation of the turbulent flow.

3.3 Determination of the Mean Reaction Rates

The mean consumption and heat release rates are deduced from the flame surface density S_f and from the rates obtained in the local flamelet analysis. One writes, for example, that the average rates per unit volume are :

$$\widetilde{W} = \bar{\rho} Q \widetilde{S}_f \ , \ \widetilde{W}_F = -\bar{\rho} \frac{Q}{(-\Delta h_F^o)} \widetilde{S}_f$$

where W and W_F are respectively the mean heat release rate per unit volume and the mean fuel consumption rate. These expressions may also be refined by taking into account the fact that the local flamelets correspond to a distribution of strain rates. It is then necessary to specify or presume a probability density function $P(\varepsilon_S)$ describing this distribution and the heat release rate becomes

$$\widetilde{W} = \overline{\rho \widetilde{S}_f} \int_0^\infty P(\varepsilon_S) Q(\varepsilon_S) d\varepsilon_S$$

One may also assume that the burnt gas temperature T_b which arises in the local flamelet model is variable and that the local reactive sheets may be characterized by a joint pdf $P(\varepsilon_S, T_b)$. Under this assumption the mean heat release takes the form

$$\widetilde{W} = \overline{\rho \widetilde{S}_f} \int P(\varepsilon_S, T_b) Q(\varepsilon_S, T_b) d\varepsilon_S dT_b$$

Expressions for the pdf may be found in Peters (1984) or Bray (1987).

An alternative procedure which accounts for the fact that the flow features a range of scales and fluctuations intensities consists of using the power spectral density of turbulence. A spectral version of the coherent flame model has been developed by us on this basis. Its advantage over the standard pdf approach is that it accounts for the diversity of scales as well as for the power contained in the turbulent eddies. Initial results obtained with this model are quite encouraging.

Up to now it has been assumed that the chemical reaction only occurs in thin flame sheets. It is however possible that at least some of the reactions will progress outside these sheets for example in well mixed relatively hot regions of the flow. Intuitively, this situation will prevail for reactions participating to the pollutant formation and destruction. It is then necessary to write additional balance equations for the species involved in these reactions, account for the mean reaction rates of the different types and specify a method which will distinguish the different combustion processes and their respective contributions.

The basic elements of the coherent flamelet model for premixed combustion are summarized in Fig. 8.

4. EXPERIMENTS ON TURBULENT DUCTED FLAMES

4.1 Experimental Set-up and Light Emission Measurement

The experimental set up used in this study is shown in Figure 9. A mixture of air and propane is injected through a long duct into a rectangular combustor. The inflow stream has a pressure of 1 bar and a temperature of 300 K. The maximum mass flow rate is 120 g/s. The height, depth and length of the combustion chamber are respectively 50, 80 and 300 mm. The upper and lower combustor walls are made of thick ceramic material while the lateral walls are transparent artificial quartz windows which allow the visualisation of the whole chamber. The inlet plane comprises a V-gutter flame holder placed at the duct center and producing a 50 % blockage. Ignition of the premixed stream is obtained with a spark plug. Combustion is then stabilised by hot gases recirculating behind the V-gutter.

In the present experiment no attempt is made to obtain a detailed description of the chemical kinetics in the reaction zone. Our scope is to get an information on the local heat release source term distribution. It is important to note that the heat release source term is far more interesting than temperature for modeling purposes. Indeed the temperature is less sensitive to the combustion model because it is integrated. On the contrary, the heat release source term is directly dependent on the combustion model and allows an interesting evaluation of it. While exact measurements of the local heat release are not available, many observations indicate that the light emitted by the combustion zone is related to the reaction intensity and hence to the heat-release process. Certain radicals like C_2 and CH appear almost exclusively in reactive zones and their concentration is always small. Hence the self absorption of the light emitted by these radicals is not important and the radiated light is directly related to the reaction rate or equivalently to the heat release rate. While a linear relation between the heat release source term and the light emission

from free radicals has been proved in some special situations it may be safely stated that a monotonic -and more or less linear- relation exists between these two quantities. The measurement of the heat release source term relies on this assumption. The calibration of the radical light emission may be obtained from an additional information (for instance gas analysis in the exit plane of the combustor) providing the global heat release in the combustion chamber.
The emission bands used to this purpose are the (0,0) C_2 and (0,0) CH bands at wavelengths of 516 and 430 nm respectively. Each wavelength is isolated in the emission spectrum with a narrowband interferential filter ($\Delta\lambda$ = 5 nm). The light emitted by the flame is collected by a f = 100 mm convex lens located at 200 mm from the combustor center plane. The light beam is detected by a photomultiplier through a 2 mm diameter pinhole to provide local measurement : the emitting area actually seen by the detector has a diameter of about 2 mm. The detector output is amplified and transmitted to a Masscomp-5600 computer. This computer also controls the optical displacement in the vertical and horizontal direction (see Figure 9). The photomultiplier scans a grid comprising 600 points (30 points horizontally and 20 vertically). At each point of the grid, 30000 data samples are acquired at a sampling frequency of 30 kHz (after a low pass 6 kHz filtering to prevent aliasing) and averaged. This process yields the spatial distribution of the mean light emission from C_2 and CH radicals.

4.2 Experimental Results : Effects of the Equivalence Ratio

As already indicated, the C_2 and CH light emission provides information on the local rate of heat release. Spatial distributions of the light emitted by CH radicals are displayed in Figures 10 . These maps correspond to the same flow rate (75 g/s) but to different values of the equivalence ratio.Measurements of the light emission of C_2 radicals yield similar patternsconfirming that C_2 and CH may be used as tracers of the local heat release in the flame. If one now examines the evolution of the mean flame structure with the equivalence ratio one observes that at low values of Φ, concentrated regions of reaction are in the vicinity of the flame holder. For $\Phi = 0.60$ these regions form a merged reaction zone. When the equivalence ratio increases, two flames develop from the lips of the flameholder. These flames become more and more separated when Φ increases. For the highest equivalence ratios, the light emission reaches its maximum in the vicinity of the two sidewalls. This peculiar phenomenon is observed because the flame now touches the boundaries which are in the present case nearly adiabatic. At the points where the flame reaches the walls, the flow in the boundary layer is decelerated and its velocity is less than that existing in the main flow and the flame angle increases (this angle is measured with respect to the axial direction). The wall temperature takes large values and the wall region acts like a secondary stabilization zone for the incoming stream of fresh mixture. In this region combustion is activated and the light emission is enhanced. For $\Phi = 0.90$ the two flames are separated by a region of reduced reaction. In each flame region the radical light emission increases progressively and reaches its maximum in the vicinity of the combustor walls. These data clearly show that the equivalence ratio significantly affects the flame structure. The modifications associated with this parameter are not extensively documented in the previous literature (see Libby et al. 1986 for a review of the available data, Zukoski and Marble 1955 for an early analysis of the influence of the mass flow rate on the wake transition behind a flame stabilizer).

5. NUMERICAL RESULTS AND COMPARISON WITH EXPERIMENTS

Numerical results corresponding to the experimental configuration are now described. The calculations are carried with a single set of constants. Standard values are retained for the k-ε model ($C_\mu = 0.09$, $C_1 = 1.44$, $C_2 = 1.92$, $\sigma_k = 1$, $\sigma_\varepsilon = 1.3$). The constants appearing in the balance of flame area are $\sigma_{S_f} = 1$, $\alpha = 10$, $\beta = 0.4$ and $C_s = 0.17$. Calculations have been carried with two distinct numerical codes :

- an iterative elliptic code (ECRIN) developped by SNECMA, based on the "SIMPLE" technique providing the steady state solution
- a time dependent finite difference algorithm providing transient as well as steady state solution (MOCROI, Dupoirieux and Scherrer 1985).

Systematic comparisons between the results obtained with these two methods show only minor differences indicating that the numerical solution is independant of the code. Tests were also carried to study grid effect and make sure that the solution did not depend on the grid selected. Conditions retained for the numerical simulation are as follows:

Code : ECRIN (iterative elliptic algorithm developped by SNECMA)
Grid : 27 vertical and 73 horizontal nodes (considering the plane of symmetry, the calculations are performed on one half of the combustion chamber). The grid is non uniform with local refinement near the stabilizer lips

Flow parameters : Mass flow rate : m = 75 g/s ,Inlet flow velocity : V =15.75 m/s, Kinetic energy at combustor inlet : k = one percent of V^2 = 2.5 m^2/s^2, Inlet turbulence lenght scale : l_T = 20 mm, Inlet turbulence dissipation : ε = 77 m^2/s^3, Inlet temperature : T = 300 K, Equivalence ratio : Φ = 0.75, Inlet fuel mass fraction : Y_{FU} = 0.0458.

Starting from steady state cold flow conditions a converged solution is obtained with ECRIN after about 700 iterations.
Some of the numerical results obtained with the model are displayed in Figures 11 to 13. Contours of constant temperature, flame surface per unit mass and mean heat release rates are plotted for two values of the equivalence ratio The mean temperature distribution appear in Fig. 11, 12 and 13-a. The temperature changes rapidly across the two reaction regions originating from the stabilizer lips. These regions may be visualized by plotting the distributions of flame area (Fig. 11,12 and 10-b) and the mean heat release terms (Fig. 11,12 and 13-c). For operating conditions corresponding to a lean mixture (Φ =0.60) the two reaction regions are nearly merged and form a pattern which resembles that observed in the experiment (Fig.10-a). As the equivalence ratio is increased (Φ=0.75) the two reaction regions shift apart and the distribution obtained is close to that found in the corresponding experiments (Figures 10-c). The agreement is not perfect but the main features are retrieved. Similar results are obtained for other flow conditions and other geometrical configurations including a multi-step dump combustor and a side-dump combustor.
As indicated before, most models provide reasonable temperature distributions but the heat release terms are not well predicted. This discrepancy is illustrated in Maistret et al. (1989) for the Eddy-Break-Up model. The regions where the heat release term reaches a maximum are very narrow. It is found in that case that the reaction rapidly goes to completion so that the maximum appears in the vicinity of the stabilizer lip. A tuning of the constants of the EBU model could of course lead to an improved pattern but the resut would have a restricted domain of validity.

6. CONFIGURATIONS COMBINING PREMIXED AND NONPREMIXED FLAMES

We have up to now considered premixed flame configurations. It appears however that in many situations premixed and non premixed flame elements may co-exist in the same flow. This is the case during the ignition phase of a cryogenic rocket engine. Combustion

in a Diesel engine also begins with the ignition of a premixed flame and ends with a diffusion flame. Another configuration of technological interest is that found in supersonic combustion devices utilizing mixing layer stabilization concepts. In such systems hydrogen is injected through a set of wedges and forms multiple mixing regions with the incoming air-stream. The structure of the flame in the vicinity of the trailing edges is determined by a competition between different physical processes (see Barrère and Mestre 1988 for a recent discussion). One mechanism of flame stabilization involves the initial mixing of the reactants (Fig. 14a). When the mixed fluid reaches a condition which allows the propagation of a premixed front then stabilization of the combustion wave takes place at the point where the flow velocity equals the normal flame speed of the mixture. Predictions of stabilization lengths based on this simple concept appear to be in good agreement with experiments carried by Baev and Yasakov (1975)
Another mechanism (Fig. 14 b) also involves an initial mixing of the reactants followed by an auto-ignition of the mixture. The distance at which ignition takes place is proportional to a local flow velocity and to a delay time accounting for the induction and heat release times. This time delay depends on the local flow variables such as the temperature and the mixture fraction and on the detailed reaction mechanism.
A third process identified by Barrère involves an initial development of patterns (pockets) of premixed gases (Fig.14 c). The temperature T_0 of these regions takes a value which lies between that of the hot air and cool hydrogen flows. Ignition then may take place at the contact surface separating the mixed fluid and the hot air-stream. In this circumstance the stabilization length is proportional to the flow velocity and to the sum of the time delays characterizing the mixing and ignition processes.

The three cases involve a form of premixed combustion which precedes the non-premixed flame and stabilizes the combustion process. One may also refer to analytical works of Linan (1988) and Dold (1987) concerning the structure of a laminar diffusion flame in the vicinity of the splitter plate separating the reactant streams. Linan indicates that under certain conditions premixed and non premixed combustion may occur simultaneously. Partially premixed diffusion laminar flames have also been computed by Rogg et al. (1986) and included in a flamelet model of a fuel jet. But, in general, the situations involving simultaneous premixed and non-premixed flames cannot be treated with classical flamelet models.
Our aim is to propose a combined flamelet representation which accounts for premixed and non premixed flame regimes. A complete presentation of this model is given in

Veynante et al. (1989)a and an application to a propagating flame problem is described in Veynante et al (1989)b.

7. COMBINED FLAMELET MODEL

For clarity the combined premixed/non premixed flamelet model is presented in its simplest form. As with any model of this kind, some aspects will need future revisions and improvements. The model will be formulated in terms of surface densities per unit volume but it is a simple matter to write a similar set of equations in terms of the surface densities per unit mass. To simplify the notations the tilde superscript is suppressed in the following development but it is recalled that the governing equations are expressed in terms of mean values of the variables.Consider a simple, one step, fast and irreversible reaction of the type

$$F + s\,O \text{ -----> } P$$

In the basic flamelet model a single flame surface density was introduced Σ_f. This quantity has the dimension of the inverse of a length (area/volume) and gives a measure of the mean flame area available at each point. In a combined flame model, two flame surface densities must be considered : one for the area of diffusion flamelets (Σ^d) , the other for the area of premixed flamelets (Σ^p). In addition we also introduce a third surface density, Σ^c, which represents a mixing area without combustion between the two non premixed reactants O and F. This area does not represent a flame surface but a disruption in a diffusion flame sheet which acts like a surface of contact. Through these "holes", the two reactants O and F may diffuse to form a premixed gas as presented in Fig. 15a and 15b. Then, in a second step (Fig. 15c), this premixed fluid may be ignited to burn as a premixed flame. This process may start by auto ignition, if the temperature is sufficiently high, spark ignition or mixing with hot gases and depends on transport and chemical kinetic mechanisms. The consumption of this local premixed gas eventually leads to a quenching of the premixed flame and serves as an ignition source to a diffusion flame between fuel and oxydizer (Fig. 15d). The last step of this mechanism may be the consumption of all premixed reactants and, under that circumstance, only a diffusion flame prevails between the unmixed oxydizer and fuel (Fig. 15e).

To be consistent with this description, it is necessary to introduce two mass fractions for each reactant i : one, Y_i^d, represents the mass of the reactant i in a non premixed form and the other, called Y_i^p, the mass of the reactant i premixed with the other reactant. From this analysis, the balance equation for the non premixed reactant may be written :

$$\frac{\partial \rho Y_i^d}{\partial t} + \frac{\partial \rho v_k Y_i^d}{\partial x_k} = \frac{\partial}{\partial x_k}\left(\frac{\mu_t}{\sigma_i^d}\frac{\partial Y_i^d}{\partial x_k}\right) - \rho V_{Di}^d \Sigma^d - \rho V_i^c \Sigma^c \tag{33}$$

where the last two terms express the consumption of reactant i in the diffusion flame sheets and the mixing of the two reactants O and F to form a premixed fluid. The consumption rate per unit of diffusion flame area, V^d_{Di}, and the mixing rate per unit of contact surface, V^c_i, are determined from local models. For the reactant i in a premixed form, the balance equation becomes

$$\frac{\partial \rho Y_i^p}{\partial t} + \frac{\partial \rho v_k Y_i^p}{\partial x_k} = \frac{\partial}{\partial x_k}\left(\frac{\mu_t}{\sigma_i^p}\frac{\partial Y_i^p}{\partial x_k}\right) - \rho V_{Di}^p \Sigma^p + \rho V_i^c \Sigma^c \tag{34}$$

where the last two terms represent respectively the consumption rate of reactant i in a premixed flame and the production of premixed fluid through the mixing surface between non premixed reactants. Of course, the balance equation for the product P of the reaction is :

$$\frac{\partial \rho Y_P}{\partial t} + \frac{\partial \rho v_k Y_P}{\partial x_k} = \frac{\partial}{\partial x_k}\left(\frac{\mu_t}{\sigma_i^p}\frac{\partial Y_P}{\partial x_k}\right) + \rho V_{Dp}^d \Sigma^d + \rho V_{Dp}^p \Sigma^p \tag{35}$$

where the two source terms describe the generation of products in diffusion and premixed flamelets

Now, in addition to the previous conservation equations three new balance relations are needed : one for each flame surface density considered. A balance equation for the diffusion flame surface density, Σ^d, may be written in the form :

$$\frac{\partial \Sigma^d}{\partial t} + v_k \frac{\partial \Sigma^d}{\partial x_k} = \frac{\partial}{\partial x_k}\left(\frac{v_t}{\sigma_\Sigma^d}\frac{\partial \Sigma^d}{\partial x_k}\right) + \alpha^d \varepsilon_S \Sigma^d - \beta^d\left(\frac{V_{Do}^d}{Y_o^d} + \frac{V_{Df}^d}{Y_f^d}\right)(\Sigma^d)^2 + R^d \tag{36}$$

where α^d and β^d are two constants and ε_S is the local strain rate.

With the exception of the last term, this equation is similar to the balance equation proposed by Marble and Broadwell for the flame surface density in a diffusion flame. This equation may be established from basic principles combined with physical modelling of the mechanism of flame surface generation and dissipation. The last term, R^d, is a production term that expresses the creation of diffusion flame surface by consumption of the premixed reactants between the non premixed reactants O and F (see Fig. 15d). This term describes the diffusion flame initiation.
One may also write a balance equation for the premixed flame surface density, Σ^p :

$$\frac{\partial \Sigma^p}{\partial t} + v_k \frac{\partial \Sigma^p}{\partial x_k} = \frac{\partial}{\partial x_k}\left(\frac{v_t}{\sigma_\Sigma^p}\frac{\partial \Sigma^p}{\partial x_k}\right) + \alpha^p \varepsilon_S \Sigma^p - \beta^p\left(\frac{V_{Do}^p}{Y_o^p} + \frac{V_{Df}^p}{Y_f^p}\right)(\Sigma^p)^2 + f_{ig} \quad (37)$$

where α^p and β^p are two new constants. The production of premixed flame surface by ignition of premixed reactants is expressed by the last term of this equation f_{ig} . This ignition function is determined from an examination ofthe local processes taking place during the flame initiation phase (auto - ignition, ...) and from the flow geometry (spark, mixing with hot gases, ...) and it may be expressed in terms of the local mass fractions of premixed reactants, temperature, stain rate, ...
The balance equation for the third surface density, Σ^c, has, logically, the same form as the two others equations :

$$\frac{\partial \Sigma^c}{\partial t} + v_k \frac{\partial \Sigma^c}{\partial x_k} = \frac{\partial}{\partial x_k}\left(\frac{v_t}{\sigma_\Sigma^c}\frac{\partial \Sigma^c}{\partial x_k}\right) + \alpha^c \varepsilon_S \Sigma^c - \beta^c\left(\frac{V_{Do}^c}{Y_o^c} + \frac{V_{Df}^c}{Y_f^c}\right)(\Sigma^c)^2 + R^c \quad (38)$$

where α^c and β^c are two additional constants.
The third term describes the mutual anihilation of contact surface due to complete mixing of the non-premixed reactants separating adjacent contact sheets. The last term R^c expresses the dissipation of the contact surface by ignition of a diffusion flame between the two non premixed reactants.
Now, we must express the two added production terms, R^d and R^c, in the balance equations for the surface density of diffusion flame and contact area. R^d is the production of surface of diffusion flame by consumption of the premixed reactants between pure

oxydizer and fuel, whereas R^c corresponds to the destruction of contact area by ignition of a diffusion flame between the two non premixed reactants. Then, logically : $R^d = -R^c$ R^d expresses the production of a diffusion flame by destruction of a premixed flame. Accordingly, R^d may be viewed as a part of the destruction term in the balance equation for the surface density of premixed flame area :

$$R^d = \gamma^d \left(\frac{V^p_{Do}}{Y^p_o} + \frac{V^p_{Df}}{Y^p_f}\right)(\Sigma^p)^2 \tag{39}$$

We now have a set of equations which simultaneously model premixed and non premixed turbulent flames. This set may be used to predict the ignition of a diffusion flame in a mixing layer. In a first step, non premixed reactants are separated by a non reactive mixing surface (Σ^c). Through this surface, fuel and oxydizer mix to form a premixed fluid. Then, when ignition occurs due to some external processes like auto - ignition, ignition by hot gases, ..., a premixed flame is formed and the corresponding flame surface density, Σ^p, is described by a balance equation. This equation includes an ignition function which models the flame initiation process. During the last step, all premixed reactants are consumed and the diffusion flame takes over. Of course, this set of equations must be supplemented with expressions for the local consumption rate of each reactant in the premixed and diffusion flame sheets and with a mixing rate per unit of contact area, $V^p{}_{Di}$, $V^d{}_{Di}$, $V^c{}_i$ respectively. In general, these consumption rates are obtained from the computations of plane laminar strained flames including complex finite rate chemistry and multiple species transport. Under the assumption of a simple, one step, fast and irreversible chemical reaction, Marble and Broadwell have derived analytical expressions for consumption rate of reactants in a diffusion flame

$$V^d_{Df} = Y^d_f(\infty)\left(\frac{D\varepsilon_S}{2\pi}\right)^{1/2}\left(\frac{\Phi+1}{\Phi}\right)e^{-w^2}$$

and

$$V^d_{Do} = Y^d_o(\infty)\left(\frac{D\varepsilon_S}{2\pi}\right)^{1/2}(\Phi+1)e^{-w^2}$$

In these expressions, Φ and w are respectively the equivalence ratio and a characteristic parameter defined by

$$\Phi = \frac{sY_f^d(\infty)}{Y_o^d(\infty)}, \quad w = erf^{-1}\left(\frac{\Phi - 1}{\Phi + 1}\right)$$

where $Y^d{}_i(\infty)$ designate the free stream mass fractions of each pure reactant .
One may similarly derive expressions for the mixing rate per unit of contact area from the calculation of the mass flow rate through a plane strained surface :

$$V_f^c = Y_f^d(\infty)\left(\frac{D\varepsilon_S}{2\pi}\right)^{1/2}, \quad V_o^c = Y_o^d(\infty)\left(\frac{D\varepsilon_S}{2\pi}\right)^{1/2}$$

The consumption rate for each reactant in the premixed flame, VP_{Di}, are function of the local strain rate, temperature, equivalence ratio, ... (see Darabiha et al. 1988), but they may be assumed to be constant for simple test cases as the one proposed below.
The basic elements of the combined flamelet model are summarized in the form of a diagram in Fig. 16.

8. APPLICATION TO A REACTIVE SHEAR LAYER

Let us consider as a test of the model, the case of a two dimensional mixing layer such as the one shown in Fig. 17. Two parallel streams, each containing one of the two reactants, interact with different free stream velocities U_1 and U_2 (the velocity ratio U_1 / U_2 is 0.2). The computation is performed with a parabolic code under the classical boundary layer assumptions and the two equations ' Saffman turbulence closure (for details about this code, see Veynante et al. 1986). The local strain rate is determined from the mean axial velocity gradient: $\varepsilon_S = \alpha\left|\partial u / \partial y\right|$.
The local consumption rates proposed in the previous section have been used for each reactant. A simplified ignition function is considered. It is assumed that ignition takes place when the mass fraction of premixed reactants reaches a certain level. The numerical values of the constants are :
$\alpha^p = \alpha^d = \alpha^c = 0.9, \; \beta^p = \beta^d = \beta^c = 0.9, \; \gamma^d = 0.1, \; \alpha = 1., \; \sigma^p = \sigma^d = \sigma^c = \sigma_p = 1.$
It is convenient to define sectional averages of the flame and contact surfaces by taking the integral of these quantities across the mixing layer at each downstream location x :

$$\overline{\Sigma}^k(x) = \frac{1}{\delta(x)}\int \Sigma^k(x,y)\, dy, \quad \text{for } k = p, d, c$$

where $\delta(x)$ is the thickness of the mixing layer at the downstream location x.

The sectional averages of each surface density (contact surface, premixed and diffusion flames) are displayed in Fig. 18 as a function of the downstream location x. . In the initial part of the mixing layer for $x \leq x_p$, the contact surface separating the two reactants grows and the mass of premixed reactants increases. Then, the premixed flame is initiated, the available contact area decreases while the premixed flame area grows ($x_p \leq x \leq x_d$), premixed reactants are consumed and reaction products are formed. In a last step, premixed reactants and premixed flamelets disappear ($x \geq x_d$) while a diffusion flame, described as a collection of laminar flame sheets, takes over. This progressive transition beetween a mode of pure mixing and a turbulent diffusion flame is in qualitative agreement with experimental observations of the near field of non premixed flames as for example those reported by Schefer et al. (1988).

Another view of this process is given by the contour plots displayed in Fig.19 and 20. Contours of constant axial velocity are plotted in Fig. 16 to show the growth and extent of the mixing layer. Contours of constant densities of contact, premixed flame and diffusion flame areas are displayed in Fig. 20 indicate how the processes of mixing, premixed and diffusion flame occur in sequence as described previously. An examination of the mass fraction distributions indicates that the premixed fuel is first formed from the initially pure reactants, products are then generated in the premixed flame region. When the premixed flamelets vanish, the generation of products proceeds in diffusion flame sheets and the spatial distribution of the product mass fraction takes a standard shape.

9. CONCLUSIONS

Flamelet models applicable to turbulent combustion are reviewed in this article. Models based on flame surface density concepts are more specifically described. The available flame surface at the point in the flow may be determined from a transport equation which accounts for production and destruction mechanisms taking place in the turbulent flowfield. The flame surface density concept may be used to analyse premixed and non premixed flames and it may be extended to other more complex configurations in which the two types of flames coexist in the same stream. Specific examples of calculations show that flamelet models based on these ideas provide viable descriptions of turbulent combustion.

ACKNOWLEDGMENTS

We wish to acknowledge the continuous support provided to this work by SNECMA. This research is also funded by DRET as a part of the combustor modeling project A3C.We also wish to adress our gratefull thanks to Professor Frank Marble for many discussions and suggestions.Thierry Poinsot is currently supported by the Center for Turbulence Research, Stanford University and by CNRS.

REFERENCES

Ashurst, W. T., Sivashinsky, G. I. and Yakhot, V. 1988, Flame Front propagation in nonsteady hydrodynamic fields, *Combustion Science and Technology*, vol.**62**, pp. 273-284.

Baev, V. K. and Yasakov, W. A. 1975, Stability of diffusion flame in a single and mixed jet, *Fissika Gorenyia i Wzryva*, vol.**11**(2), pp.

Barrere, M. and Mestre, A. 1988, Stabilisation de la flamme en combustion supersonique, *La Recherche Aérospatiale*, vol.**1**, pp. 1-13.

Borghi, R. 1985, On the structure and morphology of turbulent premixed flames. *Recent Advances in Aeronautical Science*. C. Casci and C. Bruno,ed. Pergamon, pp. 117-138.

Bray, K. N. C. 1987, Methods on including realistic chemical reaction mechanisms in turbulent combustion model. *Complex Chemical Reactions*. J. Warnatz and W. Jager,ed. Springer Verlag, Heidelberg. pp.

Bray, K. N. C., Libby, P. A. and Moss, J. B. 1984, Flamelet crossing frequencies and mean reaction rates in premixed turbulent combustion, *Combustion Science and Technology*, vol.**41**, pp. 143-172.

Brown, G.L. and Roshko, A. 1974, On density effects and large structurein turbulent mixing layers, *Journal of Fluid Mechanics* vol. 64, pp. 775-816.

Buckmaster, J. D. 1979, The quenching of two dimensional premixed flames, *Acta Astronautica*, vol.**6**, pp. 741-769.

Buckmaster, J. D. 1982, Flame stretch, *Quarterly Journal of Mechanics and Applied Mathematics*, vol.**35**, pp. 123-125.

Candel, S., Darabiha, N. and Esposito, E. 1982,Models for a turbulent premixed turbulent combustor,*AIAA 18th Joint Propulsion Conference*, Cleveland. ed.,AIAA Paper 82-1261.

Candel, S. M. and Poinsot, T. 1990, Flame stretch and the balance equation for the flame area, *to appear in Combustion Science and Technology*.

Carrier, G. F., Fendell, F. E. and Marble, F. E. 1975, The effect os strain rate on diffusion flames, *SIAM Journal of Applied Mathematics*, vol.**28**, pp. 463-500.

Clavin, P. and Williams, F. A. 1982, Effects of molecular diffusion and of thermal expansion on the structure and dynamics of premixed flames in turbulent flows of large scales and low intensity, *Journal of Fluid Mechanics*, vol.**116**, pp. 215-282.

Correa, S. M. and Shyvy, W. 1987, Computational models and methods for continuous gaseous turbulent combustion, *Progress in Energy and Combustion Science*, vol.**13**, pp. 249-292.

Darabiha, N. 1984, *Un modèle de flamme cohérente pour la combustion prémélangée. Analyse d'un foyer turbulent à élargissement brusque*, Doctoral Thesis,Ecole Centrale Paris.

Darabiha, N., Giovangigli, V., Candel, S. and Smooke, M. D. 1988, Extinction of strained premixed propane-air flames with complex chemistry, *Combustion Science and Technology*, vol.**60**, pp. 267-285.

Darabiha, N., Giovangigli, V., Trouvé, A., Candel, S. and Esposito, E. 1987, Coherent flame description of turbulent premixed ducted flames. Proceedings of the France-USA Joint Workshop on Turbulent Combustion. *Turbulent Reactive Flows*. R. Borghi and S. N. B. Murthy,ed. Springer Verlag, New York. pp. 591-637.

Djavdan, E., Darabiha, N., Giovangigli, V. and Candel, S. M. 1990, Strained propane-air flames with detailed and reduced kinetic schemes, *Combustion Science and Technology*, Submitted for publication.

Dupoirieux, F. and Scherrer, D. 1985, Methodes numériques à convergence rapide utilisées pour le calcul d'écoulements réactifs. *Numerical simulation of combustion phenomena*. R. Glowinski, B. Larrouturou and R. Temam,ed. Springer, Berlin. pp. 32-45.

Giovangigli, V. 1988, *Structure et extinction de flammes laminaires prémélangées.*, Doctorat ès Sciences.

Heitor, M.V., Taylor, A.M.K.P. and Whitelaw, J.H. 1988, Velocity and scalar characteristics of turbulent premixed flames stabilized on confined axisymmetric baffles. *Combustion Science and Technology,* vol. **62**, pp. 97-126.

Karagozian, A. R. and Marble, F. E. 1986, Study of a diffusion flame in a stretched vortex, *Combustion Science and Technology*, vol.**46**, pp. 65-84.

Kerstein, A.R., Ashurst, W.T. and Williams, F.A. 1988, Field equation for interface propagation in an unsteady homogeneous flow field. *Physical Review A*, vol. 37, pp. 2728-2731.

Lacas, F., Veynante, D. and Candel, S. M. 1989,A numerical study of propagating premixed turbulent flames,*Numerical Combustion*, Antibes. A. Dervieux and B. Larrouturou, ed.,Springer Verlag, Berlin. pp. 395-406.

Lacas, F., Zikikout, S. and Candel, S. 1987,A comparison between calculated experimental mean source terms in non premixed turbulent combustion,*AIAA/SAE/ASME 23rd Joint Propulsion Conference*, ed.,AIAA Paper 87-1782.

Laverdant, A. M. and Candel, S. 1989a, Computation of diffusion and premixed flames rolled up in vortex structures, *Journal of Propulsion and Power*, vol.**5**, pp. 134-143.

Laverdant, A. M. and Candel, S. M. 1989b, A numerical analysis of a diffusion flame vortex interaction, *Combustion Science and Technology*, vol.**60**, pp. 79.

Libby, P. A., Sivasegram, S. and Whitelaw, J. A. 1986, Premixed combustion, *Progress in Energy and Combustion Science*, vol.**12**, pp. 393-405.

Libby, P. A. and Williams, F. A., ed. 1980, *Turbulent reacting flows*. Springer Verlag,New York.

Liew, S. K., Bray, K. N. C. and Moss, J. B. 1981, A flamelet model of turbulent non premixed combustion, *Combustion Science and Technology*, vol.**27**, pp. 69-73.

Linan, A. 1988, Diffusion flame attachment and front propagation along mixing layers. *Mathematical modeling in combustion and related topics*. C. M. Brauner and C. Schmidt-Lainé,ed. Martinus Nijhoff, Dordrecht. pp. 151-154.

Liu, J. T. C. 1988,Coherent mode interaction in developing free shear layers with application to coherent flames.,Caltech, Pasadena. Marble Symposium.

Maistret, E., Darabiha, E., Poinsot, T., Veynante, D., Lacas, F., Candel, S. and Esposito, E. 1989,Recent developments in the coherent flamelet description of turbulent combustion,*Numerical Combustion*, Antibes. A. Dervieux and B. Larrouturou, ed.,Springer Verlag, Berlin. pp. 98-117.

Marble, F. E. 1985, Growth of a diffusion flame in the field of a vortex, *Advances in Aerospace Science*, , pp. 395-413.

Marble, F. E. and Broadwell, J. E. 1977, *The coherent flame model of non-premixed turbulent combustion*.Report: TRW-9-PU.

Marble, F. E. and Broadwell, J. E. 1979, *A theoretical analysis of nitric oxyde production in a methane-air turbulent diffusion flame*.EPA Technical Report, Report:

Matalon, M. 1983, On flame stretch, *Combustion Science and Technology*, vol.**31**, pp. 169-181.

Peters, N. 1984, Laminar diffusion flamelets in non-premixed turbulent combustion, *Progress in Energy and Combustion Science*, , pp. 319-339.

Peters, N. 1986,Laminar flamelet concepts in turbulent combustion,*21st Symposium (Internationa) on Combustion*, pp.

Peters, N. 1988, *Turbulent modeling, scales and flame structure of premixed turbulent flames*.Report: Technical report.

Peters, N., Göttgens, J., Klein, R. and Clavin, P. 1989*Flame surface and burning velocity in premixed turbulent flames*, vol.

Poinsot, T., Veynante, D. and Candel, S. M. 1990a,Diagrams of Premixed Turbulent Combustion Based on Direct Simulation,*submitted to 23rd Symposium (Int) on Combustion*, Orléans. ed.,The Combustion Institute.

Poinsot, T., Veynante, D. and Candel, S. M. 1990b, Direct simulation of quenching in premixed turbulent combustion, *submitted to Journal of Fluid Mechanics.*

Pope, S. B. and Cheng, W. K. 1988,The stochastic flamelet model of turbulent combustion.,*22nd Symposium (International) on Combustion*, Seattle. ed.,The Combustion Institute. pp. 781-789.

Rogg, B., Behrendt, F. and Warnatz, J. 1986,*21st Symposium (International) on Combustion*, T. C. Institute, pp.

Schefer, R. W., Namazian, N. and Kelly, J. 1988,Structural characteristicsof lifted turbulent jet flames,*22nd Symposium (Int) on Combustion*, Seattle. ed.,The Combustion Institute,Pittsburgh. pp. 833-842.

Spalding, D. B. 1978,The influence of laminar transport and chemical kinetics on the time mean reaction rate in a turbulent flame.,*17th Symposium (International) on Combustion*, Pittsburgh. T. C. Institute, pp. 431.

Strelhow, R. A. and Savage, L. D. 1978, The concept of flame stretch, *Combustion and Flame*, vol.**31**, pp. 209.

Veynante, D., Candel, S. and Martin, J. P. 1987, Coherent flame modelling of chemical reaction in a turbulent mixing layer. *Complex Chemical Reactions*. J. Warnatz and W. Jager,ed. Springer Verlag, Berlin, Heidelberg. pp. 386-398.

Veynante, D., Lacas, F. and Candel, S. 1989a,A new flamelet combustion model combining premixed and nonpremixed turbulent flames,*Aerospace Science Meeting*, ed.,AIAA Paper 89-487. pp.

Veynante, D., Lacas, F. and Candel, S. 1989b, Numerical simulation of the transient ignition regime of a turbulent diffusion flame, *submitted to AIAA journal.*

Veynante, D., Lacas, F., Maistret, E. and Candel, S. 1989c,Coherent flame model in non uniformly premixed turbulent flames,*7th Symposium on Turbulent Shear Flows*, Stanford. pp. 26.2.1-26.2.6.

Williams, F. A. 1975,A review of some theoretical combustions of turbulent flame structure,*AGARD Conference Proceedings*, pp. II.1.1-25.

Williams, F. A. 1985, *Combustion Theory* Benjamin Cummings,Menlo Park, CA.

Wright, F. H. 1959, Bluff body stabilization : blockage effects, *Combustion and Flame*, vol.**3**, pp. 319.

Zukoski, E. E. and Marble, F. E. 1955, The role of wake transition in the process of flame stabilization on bluff bodies. *Combustion Researches and Review*. Butterworths, London.

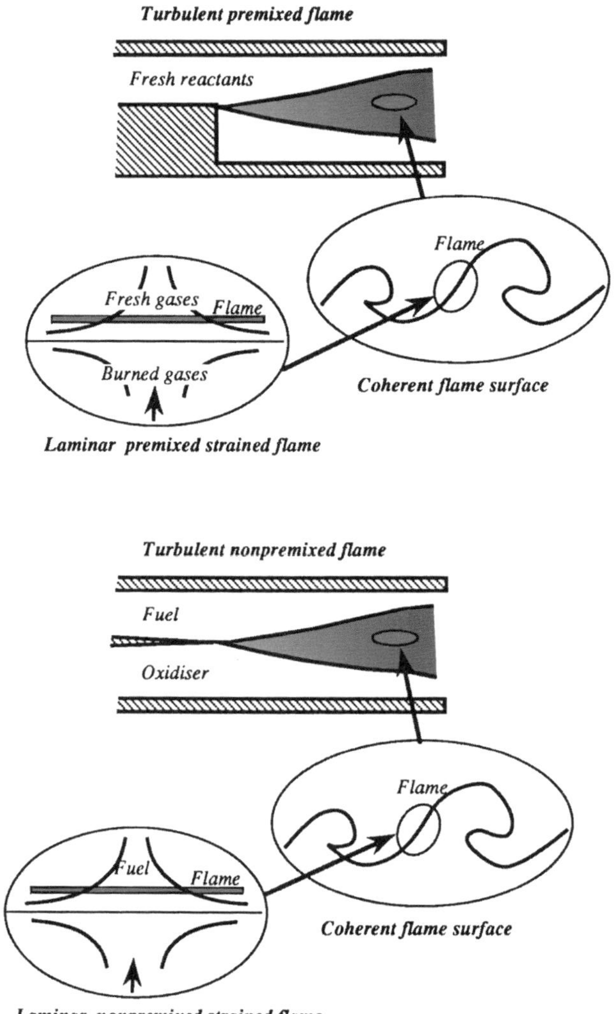

Fig. 1 Flamelet description of turbulent combustion. (a) Premixed situation (b) Non premixed situation.

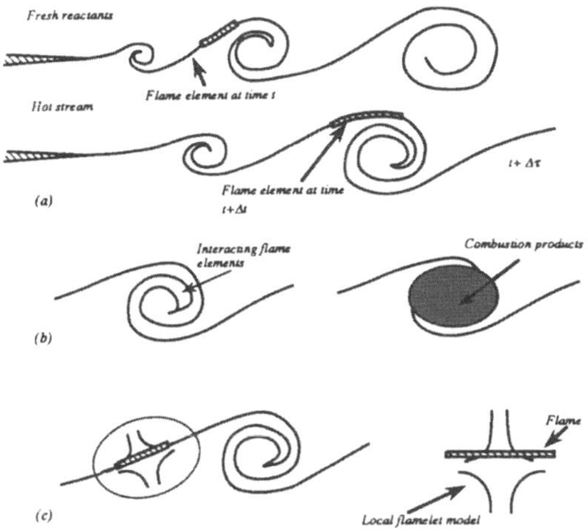

Fig. 2 Schematic representation of the fundamental processes governing turbulent combustion in the flamelet regime (a)Production of flame area by the field of strain rates (b) Mutual annihilation of adjacent flamelets (c)Strained laminar flamelet model

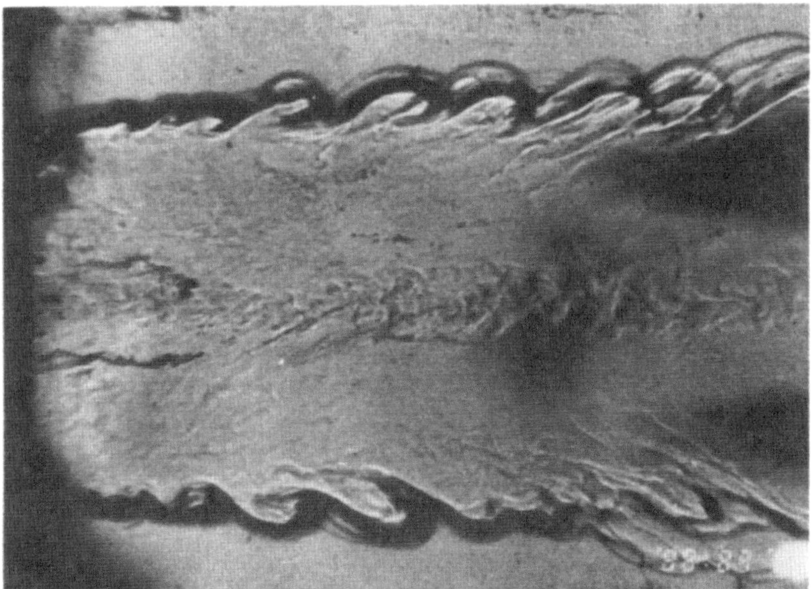

Fig. 3 Spark Schlieren photograph of a flame stabilized in a duct. The experimental configuration is described in section 4.

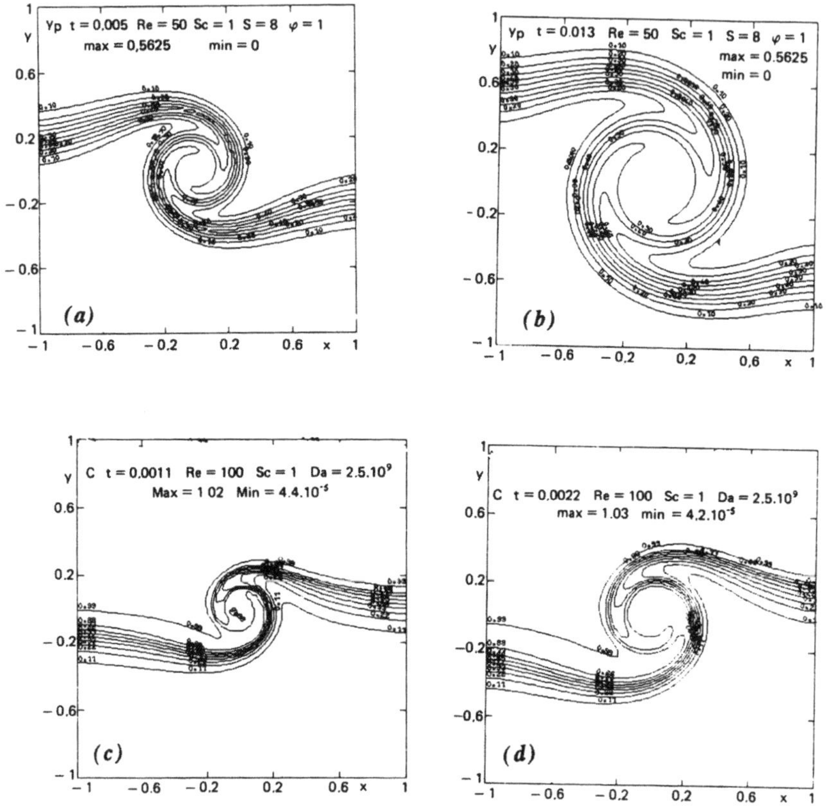

Fig. 4 Mutual annihilation of adjacent flame sheets illustrated by the interaction of a flame with a vortex. The flame is rolled up and a burnt core of products is formed near the vortex center. (a) and (b) Diffusion flame case. (c) and (d) Premixed flame case.

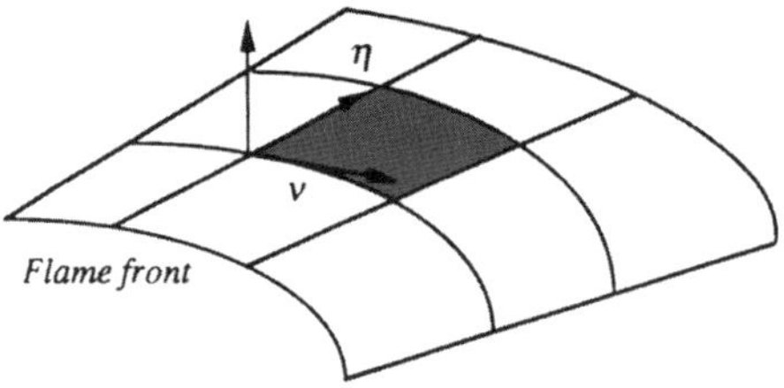

Fig. 5 A flame treated as a front in relative motion with respect to the flow.

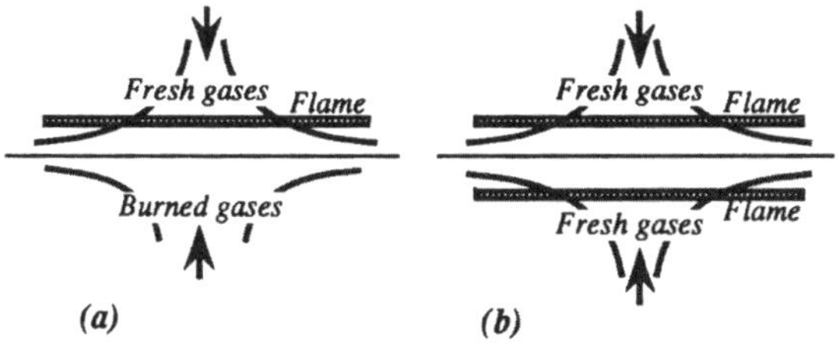

Fig. 6 Local flamelet configurations.(a)Simple flame (b)Double flame.

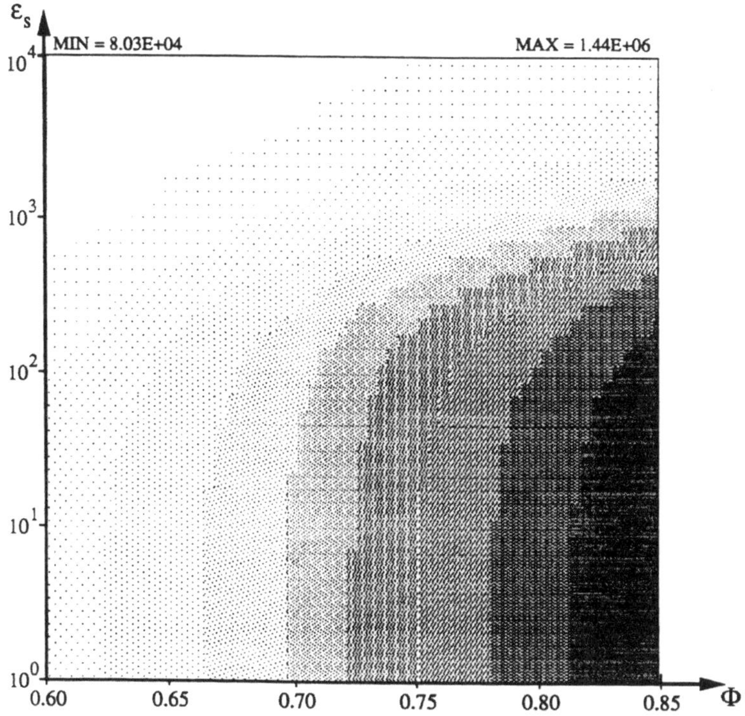

Fig.7 Heat release rate per unit flame area Q function of the equivalence ratio Φ and the strain rate ε_s. The plot is represented on a scale of grey levels. The darkest symbol corresponds to the maximum, the lightest symbol to the minimum.

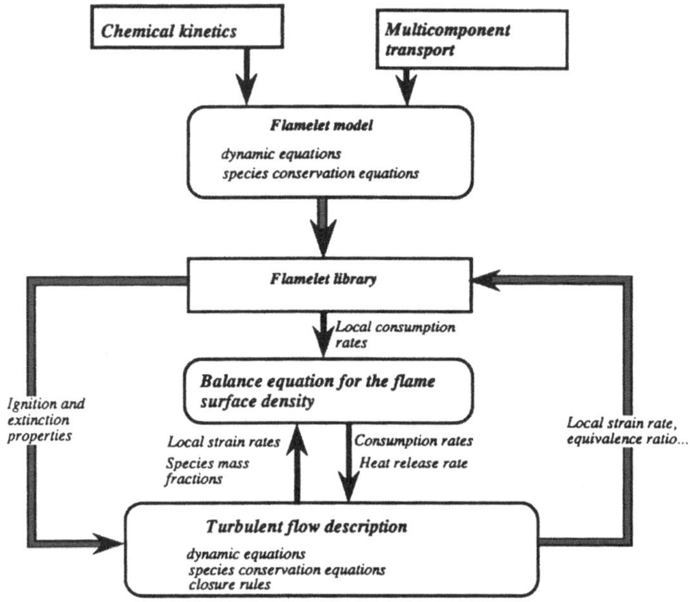

Fig. 8 Basic elements of the coherent flamelet model for premixed combustion.

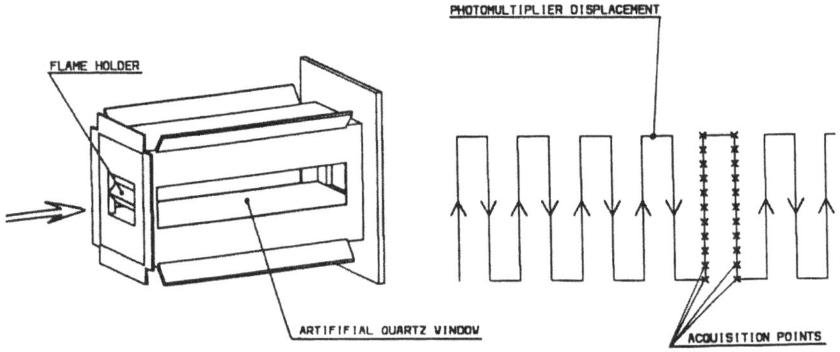

Fig.9 Experimental set-up and photmultiplier trajectory

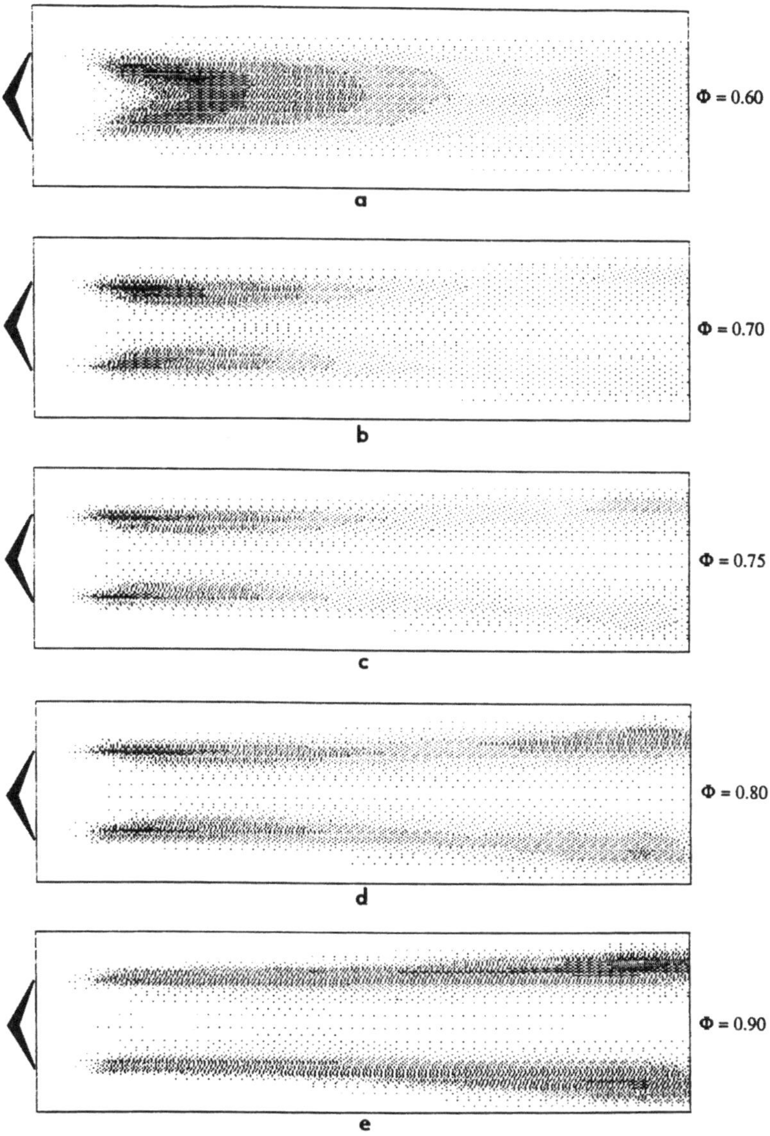

Fig. 10 Spatial distribution of the light emitted by *CH* radicals, for a mass flow rate of 75 g/s and different values of the equivalence ratio. (a) $\Phi = 0.60$, (b) $= 0.70$, (c) $\Phi = 0.75$,(d) $\Phi = 0.80$,(e) $\Phi = 0.90$.

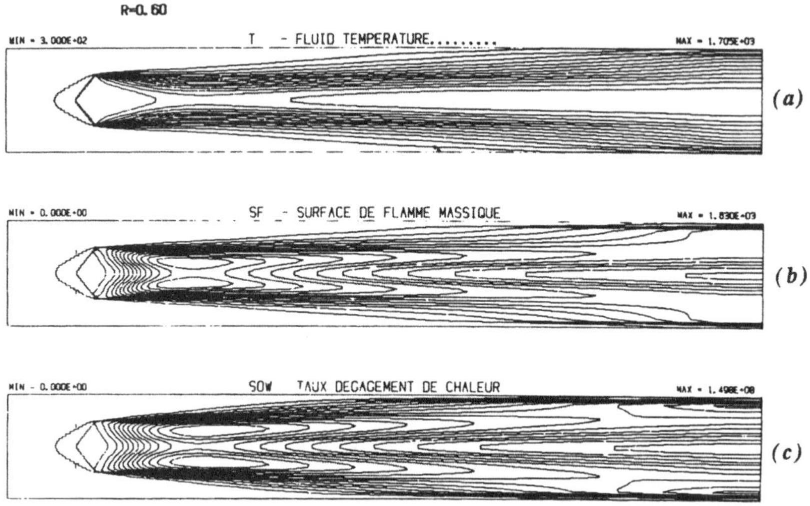

Fig. 11 Numerical calculations for a mass flow rate of 75 g/s and an equivalence ratio of 0.60 (a) Temperature (b) Flame area density (c) Heat release rate. Part "c" of this Figure may be compared to Figure 10-a

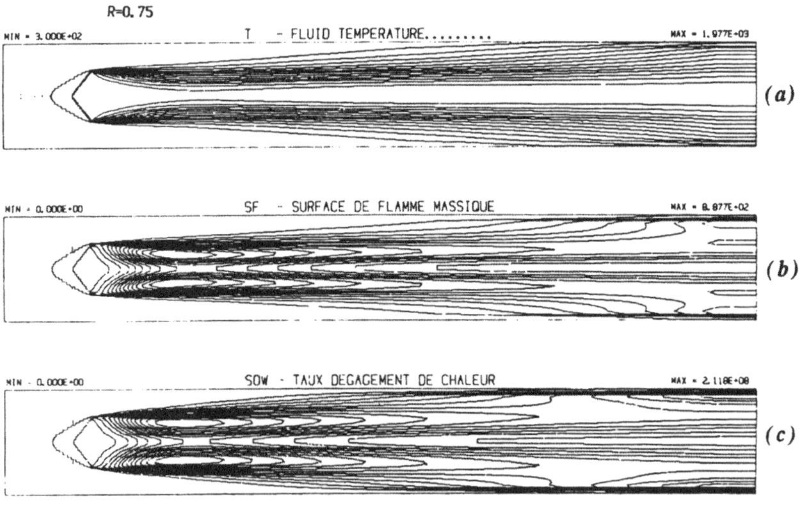

Fig. 12 Numerical calculations for a mass flow rate of 75 g/s and an equivalence ratio of 0.75 (a) Temperature (b) Flame area density (c) Heat release rate. Part "c" of this Figure may be compared to Figure 10-c

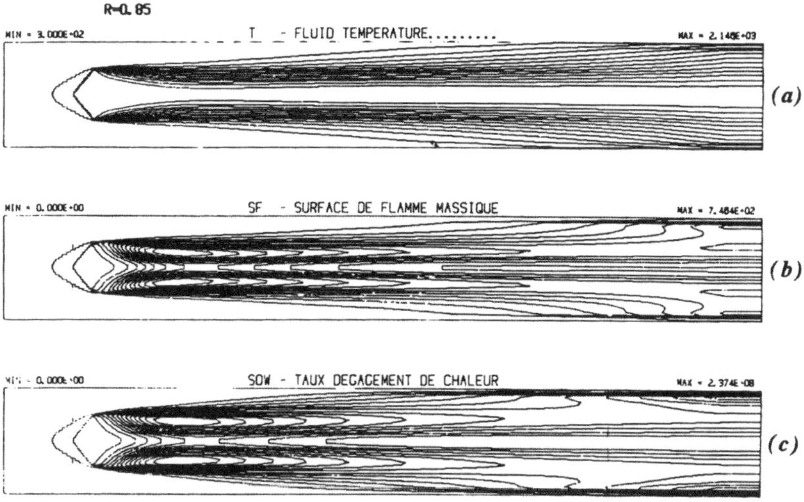

Fig. 13 Numerical calculations for a mass flow rate of 75 g/s and an equivalence ratio of 0.75 (a) Temperature (b) Flame area density (c) Heat release rate.

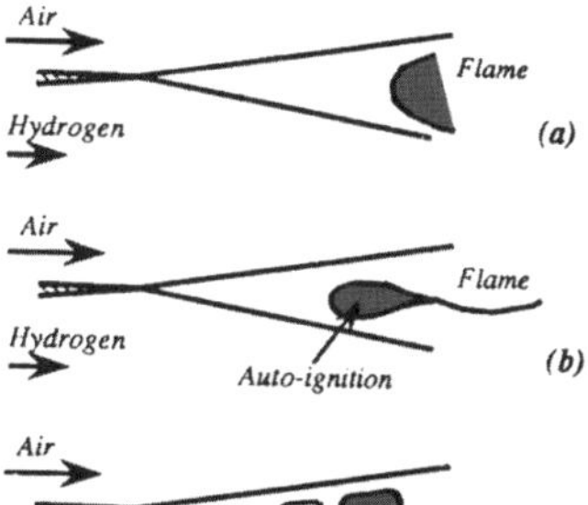

Fig.14 Flame stabilization mechanisms in supersonic combustion (according to Barrère 1988). (a) Propagation of a premixed front (b) Aut-ignition (c) Pockets of premixed gases.

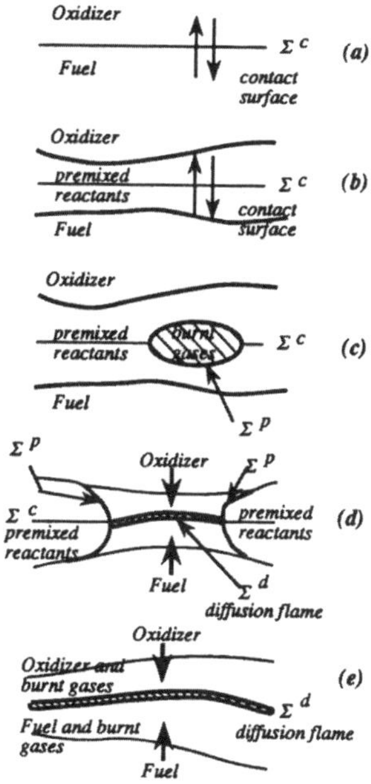

Fig. 15 Physical mechanisms described in the combined flamelet model (a) Contact between non premixed reactants (b) Production of premixed reactants by diffusion and mixing (c) Initiation and propagation of a premixed flame (d) Initiation of a diffusion flame (e) Well established diffusion flame mode.

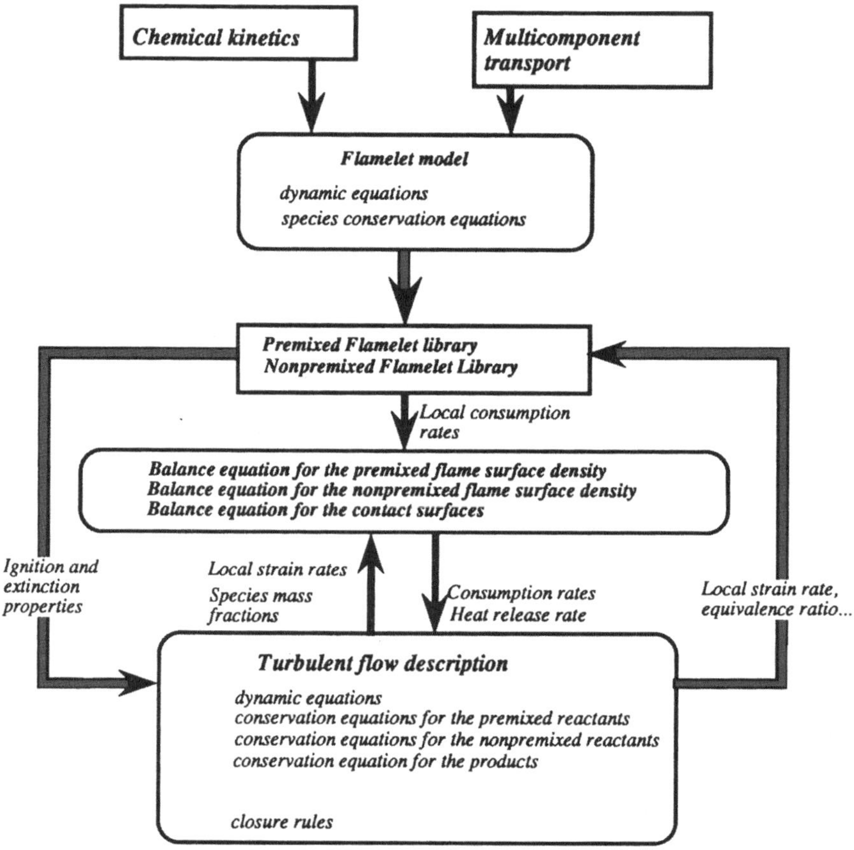

Fig. 16 Basic elements of the combined flamelet model.

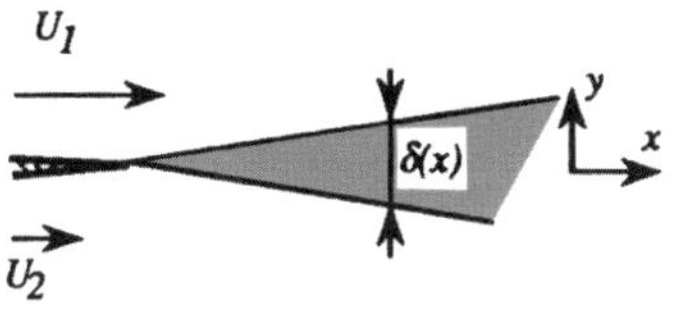

Fig. 17 A reactive mixing layer used as a test configuration.

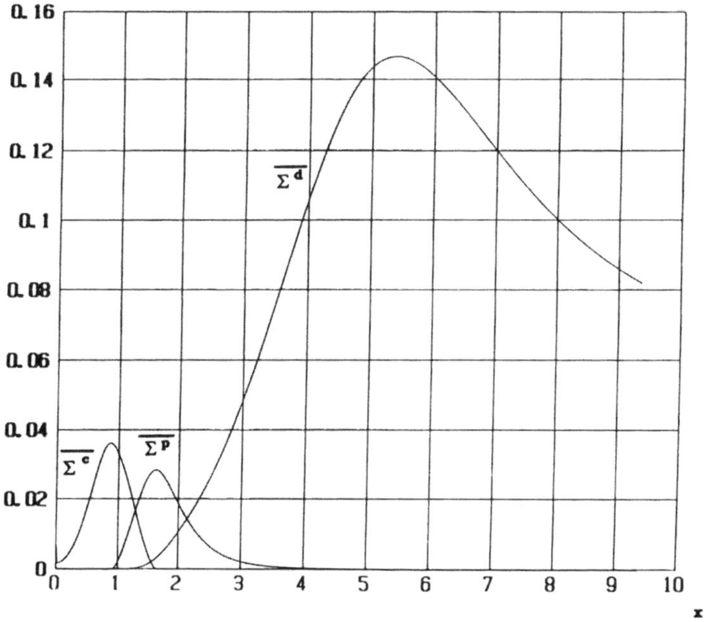

Fig. 18 Sectional averages of the surface densities.

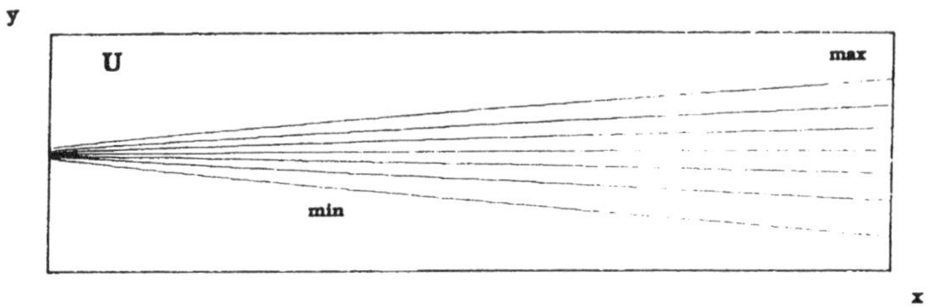

Fig. 19 Contours of constant axial velocity showing the growth of the mixing layer.

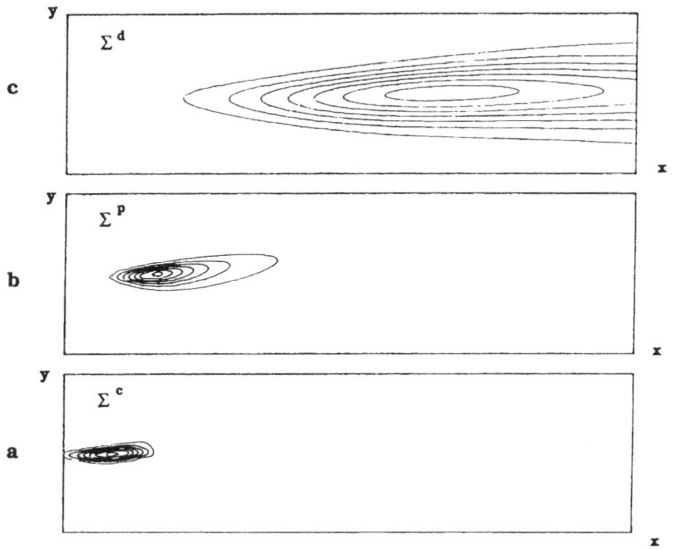

Fig. 20 Contours of constant surface densities (a) Premixed flame surface (b) Diffusion flame surface (c) Mixing surface.

MODELISATION PHYSIQUE, NUMERIQUE ET MATHEMATIQUE DES PHENOMENES DE PROPAGATION DE FLAMMES

B. LARROUTUROU

CERMICS, INRIA, Sophia-Antipolis, 06560 VALBONNE, FRANCE

RESUME: En parcourant un certain nombre de résultats significatifs tirés de la théorie des phénomènes de propagation de flammes, on décrit comment les approches physique, numérique et mathématique de la modélisation de la combustion interagissent et concourent à une meilleure compréhension de ces phénomènes.

1. INTRODUCTION

L'étude des phénomènes de combustion a connu dans ces vingt dernières années des développements très importants et attiré l'attention d'un nombre sans cesse croissant de chercheurs. Alors que les premières années de cette période de recherches intenses a surtout vu le développement des études formelles menées par des physiciens, on a également assisté plus récemment à un essor considérable des études numériques dans ce domaine, ainsi qu'au développement, moins massif mais cependant important, de travaux mathématiques.

C'est aux liens entre ces trois approches - physique, numérique et mathématique - que nous nous intéressons dans cet article. Certains de ces liens sont bien reconnus par tous: ainsi, l'importance des apports réciproques des études physiques et numériques n'est-elle pas discutée. Mais il n'en va pas toujours de même des liens avec les études mathématiques rigoureuses, dont l'intérêt est parfois mis en doute par les communautés des physiciens ou des numériciens. Nous chercherons au contraire à montrer ci-dessous combien ces trois approches se complètent et s'enrichissent mutuellement, chacune d'elle contribuant de façon intéressante et spécifique au progrès des connaisances qui sont aujourd'hui les nôtres en théorie de la combustion.

Cette étude nous donnera l'occasion de passer en revue un certain nombre de résultats physiques, numériques ou mathématiques, récents pour la plupart. Nous nous limiterons dans cette revue (qui n'a aucune prétention à l'exhaustivité) à la théorie des phénomènes de propagation de flammes, c'est-à-dire à des problèmes fondamentaux comme l'existence, l'unicité ou la multiplicité et la stabilité des solutions des modèles de base de la théorie de la combustion, laissant de côté ici les problèmes moins académiques.

Avant de présenter (dans cet ordre, dans les Sections 3, 4 et 5) différents exemples de résultats physiques, numériques et mathématiques dont l'apport pour les autres approches est important, nous rappelons dans la Section 2 les principaux modèles mathématiques qui seront utiles plus loin.

2. MODELES PHYSIQUES

Nous présentons dans cette partie les principaux systèmes d'équations utilisés en théorie de la combustion pour décrire la propagation d'une flamme prémélangée, et qui sont l'objet des études physiques, mathématiques ou numériques considérées plus loin.

Dans toute la suite, nous considérerons une flamme se propageant dans un mélange gazeux constitué de deux espèces $\mathcal{R}$ et $\mathcal{P}$ seulement, susceptibles de donner lieu à la réaction chimique simple $\mathcal{R} \rightarrow \mathcal{P}$, dans un tube cylindrique infini S.

Une grande partie des résultats présentés ci-après concernent la situation où la flamme se propageant dans le tube S est plane – il s'agit du problème de base de la théorie de la combustion. Nous commençons donc par écrire les différents systèmes d'équations utilisés pour décrire la propagation de la flamme plane. Nous présenterons ensuite le modèle auquel nous nous intéresserons pour l'étude des flammes plissées. Enfin, nous écrirons aussi le modèle classique de propagation de flammes en présence de pertes de chaleur.

2.1. Les équations d'un écoulement réactif compressible plan

Sous les hypothèses simplificatrices habituelles (gaz parfaits, chaleurs massiques C_p et C_v identiques pour les deux espèces et constantes, loi de Dalton, flux de diffusion moléculaire donné par la loi de Fick, flux conductif de chaleur donné par la loi de Fourier), et en négligeant pour simplifier les effets visqueux, on décrit classiquement la propagation de la flamme par le système suivant:

$$\begin{cases} \rho_\tau + (\rho u)_\xi = 0 \ , \\ (\rho u)_\tau + (\rho u^2 + p)_\xi = 0 \ , \\ E_\tau + [u(E+p)]_\xi = (\lambda T_\xi)_\xi + Q(\rho D Y_\xi)_\xi \ , \\ (\rho Y)_\tau + (\rho u Y)_\xi = -m\omega + (\rho D Y_\xi)_\xi \ . \end{cases} \tag{2.1}$$

Ces quatre équations traduisent la conservation de la masse, de la quantité de mouvement et de l'énergie pour le mélange, et la conservation de la masse de l'espèce réactante. Les notations utilisées sont classiques: ξ et τ sont les coordonnées d'espace et de temps, ρ est la masse volumique du mélange, u est la vitesse du mélange suivant la direction ξ (l'hypothèse de flamme plane signifie que toutes les variables ne dépendent pas des coordonnées spatiales transversales, et que les composantes de la vitesse du mélange suivant ces directions transversales à la direction ξ sont nulles), p est la pression du mélange, E est l'énergie totale du mélange par unité de volume, T est la température du mélange et Y est la fraction massique de l'espèce réactante. De plus, λ est la conductivité thermique du mélange, D est le coefficient de diffusion moléculaire, Q est la quantité de chaleur dégagée par la réaction chimique exothermique $\mathcal{R} \to \mathcal{P}$ par unité de masse du réactant, m est la masse molaire des deux espèces en présence, et R est la constante des gaz. Enfin, ω est le taux d'avancement de la réaction chimique.

Le système (2.1) est complété par les relations suivantes, issues des hypothèses rappelées plus haut:

$$mp = \rho R T \ , \tag{2.2}$$

$$E = \rho Y(h^0_{\mathcal{R}} + C_v T) + \rho(1-Y)(h^0_{\mathcal{P}} + C_v T) + \frac{1}{2}\rho u^2 = Q\rho Y + \rho C_v T + \rho h^0_{\mathcal{P}} + \frac{1}{2}\rho u^2 \ . \tag{2.3}$$

On note ici $h^0_{\mathcal{R}}$ et $h^0_{\mathcal{P}}$ les enthalpies de formations des deux espèces.

Enfin, le taux de réaction chimique est donné par la loi d'action de masse et la loi d'Arrhenius et s'écrit:

$$\omega = \omega(\rho Y, T) = \frac{\rho Y}{m} F(T) = \mathcal{B}\frac{\rho Y}{m} T^r \exp\left(-\frac{\mathcal{E}}{RT}\right) \ , \tag{2.4}$$

où $\mathcal{B}$, r et $\mathcal{E}$ sont trois constantes positives: $\mathcal{B}$ est appelé le préfacteur d'Arrhenius, et $\mathcal{E}$ est l'énergie d'activation de la réaction chimique.

L'établissement des équations précédentes à partir des théories de base de la physique (théorie cinétique des gaz et mécanique des milieux continus notamment) est présenté par exemple dans Buckmaster-Ludford [24], Larrouturou [58], et Williams [88].

Nous ne préciserons pas ici quelles conditions aux limites ou initiales sont associées au système (2.1)-(2.4), et nous passons maintenant à une approximation de ce système très utilisée en combustion, l'approximation isobare.

2.2. Les équations des flammes planes isobares

La principale approximation utilisée pour l'étude des phénomènes de propagation de flammes provient du caractère fortement subsonique de ces phénomènes: la valeur d'un nombre de Mach M représentatif de l'écoulement est de l'ordre de 10^{-2} ou 10^{-3}. Un développement formel des variables en fonction des puissances du petit paramètre M montre alors que la pression du mélange est constante en espace à l'ordre M^2, c'est-à-dire que l'on peut écrire:

$$p(\xi,\tau) = P(\tau) + \tilde{p}(\xi,\tau) \ , \tag{2.5}$$

avec $\dfrac{\tilde{p}}{P} = O(M^2) \ll 1$ (voir par exemple Joulin [45], Larrouturou [58], Majda [63]).

Dans le cadre de cette approximation, le système (2.1) devient (voir par exemple Buckmaster-Ludford [24], Larrouturou [58] pour les détails):

$$\begin{cases} \rho_\tau + (\rho u)_\xi = 0 \ , \\ \rho u_\tau + \rho u u_\xi = -\tilde{p}_\xi \ , \\ \rho C_p T_\tau + \rho u C_p T_\xi = (\lambda T_\xi)_\xi + mQ\omega(\rho Y, T) + P_\tau \ , \\ \rho Y_\tau + \rho u Y_\xi = -m\omega(\rho Y, T) + (\rho D Y_\xi)_\xi \ , \\ \\ \rho R T = mP(\tau) \ , \quad \omega(\rho Y, T) = \dfrac{\rho Y}{m} F(T) \ . \end{cases} \tag{2.6}$$

Dans (2.6), nous avons écrit les équations sous forme non conservative, et nous avons écrit l'équation de conservation de l'énergie en fonction de la température. Pour former un problème bien posé, ce système doit être complété par des conditions aux limites ou initiales. C'est ce que nous décrivons maintenant, pour des solutions instationnaires de (2.6), puis pour des solutions stationnaires.

2.2.1. Les équations des flammes planes isobares instationnaires

Il est bien connu que l'étude des solutions instationnaires de (2.6) est plus simple si l'on utilise au lieu de (2.6) une description lagrangienne de la propagation de la flamme. On introduit donc le changement de variables $(\xi,\tau) \longleftrightarrow (x,t)$ où $x(\xi,\tau)$ représente la coordonnée lagrangienne de la particule située à l'abcisse ξ à l'instant τ (inversement, $\xi(x,t)$ est la position au temps t de la particule de coordonnée lagrangienne x), et où $t=\tau$. Ce changement de variables est défini par la relation:

$$\int_{\xi(0,t)}^{\xi(x,t)} \rho(\xi',\tau)d\xi' = x \ , \tag{2.7}$$

qui implique $x_\xi = \rho$, $x_\tau = -\rho u$ (nous renvoyons par exemple à Larrouturou [56], [58] pour plus de détails sur l'introduction de la coordonnée lagrangienne x). Pour toute variable w, on a alors:

$$w_\tau = w_t - \rho u w_x \ , \quad w_\xi = \rho w_x \ , \tag{2.8}$$

et la forme lagrangienne de (2.6) s'écrit alors:

$$\begin{cases} \rho_t + \rho^2 u_x = 0 \ , \\ u_t + \tilde{p}_x = 0 \ , \\ T_t = \dfrac{mQ}{C_p}\dfrac{\omega(\rho Y,T)}{\rho} + \dfrac{1}{C_p}(\lambda\rho T_x)_x + \dfrac{1}{\rho C_p}P'(t) \ , \\ Y_t = -m\dfrac{\omega(\rho Y,T)}{\rho} + (\rho^2 D Y_x)_x \ , \\ \rho RT = mP(t) \ , \quad \omega(\rho Y,T) = \dfrac{\rho Y}{m}F_\epsilon(T) \ . \end{cases} \tag{2.9}$$

Nous avons utilisé la notation F_ϵ au lieu de F pour la partie non linéaire du taux de réaction de façon à mettre en évidence que ce terme dépend de l'énergie d'activation: ici et ci-dessous, ϵ est un paramètre positif petit, proportionnel à l'inverse de l'énergie d'activation $\mathcal{E}$.

En supposant que le nombre de Lewis $\mathcal{L} = \dfrac{\lambda}{\rho C_p D}$ est constant, nous pouvons réécrire (2.9) sous la forme:

$$\begin{cases} T_t = \dfrac{Q}{C_p}YF_\epsilon(T) + \dfrac{1}{C_p}(\lambda\rho T_x)_x + \dfrac{1}{\rho C_p}P'(t) \ , \\ Y_t = -YF_\epsilon(T) + \dfrac{1}{C_p\mathcal{L}}(\lambda\rho Y_x)_x \ , \end{cases} \tag{2.10}$$

$$\begin{cases} \rho RT = mP(t) \ , \\ u_x = \left(\dfrac{1}{\rho}\right)_t \ , \\ \tilde{p}_x = -u_t \ . \end{cases} \tag{2.11}$$

Ainsi, l'utilisation de la coordonnée lagrangienne x découple le système de réaction-diffusion (2.10) pour les "variables de combustion" T et Y du système linéaire (2.11) pour les "variables hydrodynamiques" ρ, u et $\tilde{p}$.

La forme des équations (2.10)-(2.11) suggère aussi quelles conditions initiales ou aux limites doivent être ajoutées pour obtenir un problème bien posé, Pour le système parabolique (2.10), il faut une condition initiale et deux conditions aux limites pour chacune des variables T et Y. Ecrivons donc:

$$T(x,0) = T^0(x) \ , \quad Y(x,0) = Y^0(x) \ , \tag{2.12}$$

$$\begin{cases} T(-\infty,t) = T_u(t) \ , \quad Y(-\infty,t) = Y_u(t) \ , \\ T(+\infty,t) = T_b(t) \ , \quad Y(+\infty,t) = Y_b(t) \ , \end{cases} \tag{2.13}$$

où les fonctions $T_u(t)$, $Y_u(t)$, $T_b(t)$, $Y_b(t)$ sont définies par le système différentiel:

$$\begin{cases} T_u'(t) = \dfrac{Q}{C_p} Y_u(t) F_\epsilon[T_u(t)] + \dfrac{1}{\rho_u(t) C_p} P'(t) \ , \quad T_u(0) = T_{u0} \ , \\ T_b'(t) = \dfrac{Q}{C_p} Y_b(t) F_\epsilon[T_b(t)] + \dfrac{1}{\rho_b(t) C_p} P'(t) \ , \quad T_b(0) = T_{b0} \ , \\ Y_u'(t) = -Y_u(t) F_\epsilon[T_u(t)] \ , \quad Y_u(0) = Y_{u0} \ , \\ Y_b'(t) = -Y_b(t) F_\epsilon[T_b(t)] \ , \quad Y_b(0) = Y_{b0} \ ; \end{cases} \tag{2.14}$$

dans (2.14), nous avons posé $T_u^0 = T^0(-\infty)$, $T_b^0 = T^0(+\infty)$, $Y_u^0 = Y^0(-\infty)$, $Y_b^0 = Y^0(+\infty)$, et $\rho_u(t) = \rho(-\infty,t)$, $\rho_b(t) = \rho(+\infty,t)$. Bien sûr, nous avons déduit le système différentiel (2.14) de (2.10).

De plus, puisque la deuxième et la troisième équation de (2.11) donnent u_x et $\tilde{p}_x$, une seule condition aux limites suffira pour la vitesse et pour la pression:

$$u(-\infty,t) = u_u \ , \quad \tilde{p}(-\infty,t) = p_u(t) \ . \tag{2.15}$$

(la valeur u_u ne dépend pas du temps d'après (2.11.c) puisque $\tilde{p}_x(-\infty,t) = 0$; par contre, $p_u(t)$ est une fonction donnée du temps).

Enfin, il nous faut une condition supplémentaire qui fixera l'évolution de la pression $P(t)$. Puisque $mP(t) = \rho_u(t) RT_u(t)$, il suffit de donner la valeur de $\rho_u(t)$.

Comme l'on doit avoir $u_x(-\infty, t) = 0$ d'après (2.15), on voit d'après (2.11) que $\rho_u(t)$ ne doit pas varier. Nous écrivons donc:

$$\rho(-\infty, t) = \rho_u \ , \tag{2.16}$$

une constante donnée.

Un cas particulier du problème (2.10)-(2.16) concerne le cas où les conditions aux limites (2.13) et la pression P ne dépendent pas du temps. Nous considèrerons aussi cette situation dans la Section 5.1.2 ci-dessous.

Terminons ce paragraphe en réécrivant (2.10)-(2.16) sous une forme normalisée plus simple. Nous supposerons, uniquement pour simplifier l'exposé, que le produit $\lambda\rho$ et le nombre de Lewis $\mathcal{L}$ sont constants, hypothèse qui rend constant les coefficients de diffusion dans (2.10). On peut alors choisir les valeurs de référence pour x, t, T, Y, P, ρ, u et $\tilde{p}$ pour que, en variables normalisées, le problème s'écrive (voir Berestycki *et al.* [18], Roquejoffre [76]):

$$\begin{cases} T_t = Y f_\epsilon(T) + T_{xx} + \dfrac{\gamma - 1}{\gamma} \dfrac{T_u'(t)}{T_u(t)} T \ , \\[2ex] Y_t = -Y f_\epsilon(T) + \dfrac{1}{\mathcal{L}} Y_{xx} \ , \end{cases} \tag{2.17}$$

$$T(x,0) = T^0(x) \ , \quad Y(x,0) = Y^0(x) \ , \tag{2.18}$$

$$\begin{cases} T(-\infty,t) = T_u(t) \ , \quad Y(-\infty,t) = Y_u(t) \ , \\ T(+\infty,t) = T_b(t) \ , \quad Y(+\infty,t) = Y_b(t) \ , \end{cases} \tag{2.19}$$

$$\begin{cases} T_u'(t) = \gamma Y_u(t) f_\epsilon[T_u(t)] \ , \quad T_u(0) = T_u^0 = T^0(-\infty) \ , \\[1ex] T_b'(t) = Y_b(t) f_\epsilon[T_b(t)] + \dfrac{\gamma - 1}{\gamma} \dfrac{T_u'(t)}{T_u(t)} T_b(t) \ , \quad T_b(0) = T_b^0 = T^0(+\infty) \ , \\[1ex] Y_u'(t) = -Y_u(t) f_\epsilon[T_u(t)] \ , \quad Y_u(0) = Y_u^0 = Y^0(-\infty) \ , \\[1ex] Y_b'(t) = -Y_b(t) f_\epsilon[T_b(t)] \ , \quad Y_b(0) = Y_b^0 = Y^0(+\infty) \ , \end{cases} \tag{2.20}$$

$$\begin{cases} \rho T = T_u(t) \ , \\[1ex] \rho_t + \rho^2 u_x = 0 \ , \\ \tilde{p}_x = -u_t \ , \end{cases} \tag{2.21}$$

$$\begin{cases} \rho(-\infty,t) = 1 \ , \\ u(-\infty,t) = u_u \ , \\ \tilde{p}(-\infty,t) = p_u(t) \ . \end{cases} \tag{2.22}$$

Nous renvoyons à [76] pour les détails concernant cette normalisation. Signalons simplement ici que le terme non linéaire f_ϵ est proportionnel à F_ϵ, et que le rapport $\frac{\gamma-1}{\gamma}$ dans (2.19) et (2.22) vient de la relation de Mayer $\frac{\gamma-1}{\gamma} = \frac{R}{mC_p}$, où γ est le rapport des chaleurs specifiques du mélange: $\gamma = \frac{C_p}{C_v}$.

2.2.2. Les équations des flammes planes isobares stationnaires

Venons en maintenant aux solutions stationnaires de (2.6).

Pour une solution stationnaire, la première équation (2.6.a) s'écrit:

$$(\rho u)_x = 0 , \tag{2.23}$$

d'où on déduit $\rho u = c$, une constante inconnue. Il est alors clair qu'une solution stationnaire en coordonnées eulériennes correspond à une solution d'onde progressive en variables lagrangiennes, puisque les identités $T_\tau = 0$, $Y_\tau = 0$ impliquent alors $T_t - cT_x = 0$, $Y_t - cY_x = 0$ d'après (2.8) et (2.23).

Considérons donc les solutions d'onde progressive $T(x+ct)$, $Y(x+ct)$ en variables lagrangiennes. Avec la même normalisation que précédemment, nous pouvons écrire le système vérifié par ces solutions:

$$\begin{cases} cT_x = Yf_\epsilon(T) + T_{xx} , \\ cY_x = -Yf_\epsilon(T) + \frac{1}{\mathcal{L}}Y_{xx} . \end{cases} \tag{2.24}$$

Remarque 1: Nous étudierons seulement le système (2.24) qui détermine T, Y *et le réel* c, puisque résoudre ensuite le système linéaire donnant les variables ρ, u et $\tilde{p}$ ne pose aucune difficulté. •

Remarque 2: Nous aurions pu obtenir les équations (2.24) avec la coordonnée d'espace eulérienne ξ au lieu de x si nous avions supposé la conductivité thermique λ constante. Sans hypothèse sur λ, on peut encore définir un changement de variable $\xi \longleftrightarrow x$ qui permette de retrouver le système (2.24) pour les solutions stationnaires de (2.6) (voir par exemple [58]). •

A (2.24) sont associées les conditions aux limites suivantes:

$$\begin{cases} T(-\infty) = T_u^0 , \quad Y(-\infty) = Y_u^0 , \\ T(+\infty) = T_b^0 , \quad Y(+\infty) = Y_b^0 . \end{cases} \tag{2.25}$$

Nous supposerons pour l'étude des flammes stationnaires que les données T_u^0, T_b^0, Y_u^0, et Y_b^0 satisfont:

$$T_b^0 = T_u^0 + Y_u^0 \ , \quad Y_u^0 > 0 \ , \quad Y_b^0 = 0 \ . \tag{2.26}$$

Dans ces relations, T_u^0 et T_b^0 désignent respectivement les températures normalisées du mélange frais et des gaz brûlés, et Y_u^0 est la fraction massique normalisée du réactant dans le mélange frais. La dernière égalité (2.26) exprime le fait que le réactant est complètement consommé par la flamme; l'inégalité dans (2.26) exprime que le mélange frais doit réellement contenir du réactant. Enfin, la première relation dans (2.26), qui montre que les gaz brûlés sont plus chauds que les gaz frais, traduit un bilan obtenu en intégrant de $-\infty$ à $+\infty$ la somme des deux équations (2.24).

Le problème (2.24)-(2.25) (ou le problème plus simple que l'on obtient lorsque le nombre de Lewis est égal à 1) a été étudié par de nombreux auteurs (voir par exemple Aronson-Weinberger [4], Johnson [43], Johnson-Nachbar [44], Kanel' [47], [48], Kolmogorov *et al.* [52], Zeldovich [89], Zeldovich *et al.* [90], [91], et plus récemment Berestycki *et al.* [20], Marion [65], [66], Berestycki-Larrouturou [13], [14]). Il soulève la fameuse difficulté de la frontière froide: il est facile de voir que, tel quel, le problème (2.24)-(2.26) n'a pas de solution, parce que le taux de réaction n'est pas nul dans le mélange frais: $Y_u^0 f_\epsilon(T_u^0) \neq 0$ (le fait que $\lim\limits_{x\to-\infty} [cT_x - T_{xx}] \neq 0$ contredit le fait que T est bornée au voisinage de $-\infty$). D'un point de vue physique, l'origine de la difficulté est claire: l'état (Y_u^0, T_u^0) prescrit à $-\infty$ n'est pas un état d'équilibre, ce qui entraine que le problème est mal posé.

Une manière classique de remédier à cette difficulté de la frontière froide (que nous discuterons plus longuement dans la Section 5.1.2 ci-dessous) consiste à modifier le terme de réaction en introduisant une température d'ignition. Introduisons en même temps une autre normalisation très classique du problème (2.24)-(2.25). En utilisant les variables $\Theta = \dfrac{T - T_u^0}{T_b^0 - T_u^0}$ et $Z = \dfrac{Y}{Y_u^0}$, et en introduisant une fonction f_ϵ^0 telle que:

$$\exists \theta \in]0,1[\ , \quad f_\epsilon^0 \equiv 0 \text{ sur } [0,\theta] \ , \quad f_\epsilon^0 > 0 \text{ sur }]\theta, 1] \ , \tag{2.27}$$

où θ est la température d'ignition normalisée, on cherche alors $\Theta(x)$, $Z(x)$ et c satisfaisant:

$$\begin{cases} c\Theta_x = Z f_\epsilon^0(\Theta) + \Theta_{xx} \ , \\ cZ_x = -Z f_\epsilon^0(\Theta) + \dfrac{1}{\mathcal{L}} Z_{xx} \ , \end{cases} \tag{2.28}$$

et les conditions aux limites:

$$\begin{cases} \Theta(-\infty) = 0 \ , & Z(-\infty) = 1 \ , \\ \Theta(+\infty) = 1 \ , & Z(+\infty) = 0 \ . \end{cases} \tag{2.29}$$

Rappelons enfin que ce problème se simplifie encore dans le cas où $\mathcal{L} = 1$, puisque (2.28) et (2.29) impliquent alors que $\Theta + Z \equiv 1$. Le problème se réduit alors à une seule équation différentielle: il reste à trouver une fonction $\Theta(x)$ et un réel c tels que:

$$\begin{cases} c\Theta_x = g_\epsilon^0(\Theta) + \Theta_{xx} \ , \\ \Theta(-\infty) = 0 \ , \quad \Theta(+\infty) = 1 \ ; \end{cases} \tag{2.30}$$

nous avons posé ici $g_\epsilon^0(s) = (1-s)f_\epsilon^0(s)$.

2.3. Le modèle thermo-diffusif multi-dimensionnel

Nous venons de voir que, dans le cadre de l'approximation isobare et dans une géométrie monodimensionnelle, un découplage se produit entre les "variables de combustion" T et Y (ou Θ et Z) et les variables hydrodynamiques ρ, u et $\tilde{p}$. Dans un cadre multi-dimensionnel, ce découplage n'apparaît pas: même dans le cadre de l'approximation isobare, où les effets acoustiques n'apparaissent plus, les effets diffusifs, réactifs et hydrodynamiques (convection, expansion des gaz...) présents dans le phénomène restent étroitement couplés, de façon non linéaire. C'est la raison pour laquelle les physiciens ont introduit une simplification supplémentaire, *l'approximation de densité constante* (voir Barenblatt *et al.* [5], Sivashinsky [82], [83]). Cette approximation conduit dans une géométrie multi-dimensionnelle à un système d'équations plus simple, dit thermo-diffusif, dans lequel apparaît à nouveau un découplage, et qui a permis des progrès importants dans la compréhension des phénomènes de propagation de flammes (voir [82], [83]). Dans le cadre de cette approximation, la combustion n'a pas d'effet sur le champ de vitesse, qui est donné et satisfait les équations des écoulements incompressibles.

Pour le cas qui nous intéresse où la combustion a lieu dans un tube cylindrique infini S, il est classique de considérer un champ de vitesse parallèle aux parois du tube. En notant le tube $S = \{(\xi, \bar{\eta}) \in I\!R^3,\ \xi \in I\!R,\ \bar{\eta} \in \mathcal{O}\}$, où $\mathcal{O}$ désigne un domaine borné non vide du plan, et le vecteur vitesse $\vec{V} = (u, v, w)$, on suppose alors que $v \equiv 0$, $w \equiv 0$; les équations des écoulements incompressibles impliquent ensuite $u = u(\bar{\eta})$, et il reste pour Θ et Z les équations suivantes (que nous écrivons sous forme normalisée analogue à (2.28) et dans le cas où les conditions aux limites ne dépendent pas du temps):

$$\begin{cases} \Theta_\tau + u(\bar{\eta})\Theta_\xi = Zf_\epsilon(\Theta) + \Delta\Theta \ , \\ Z_\tau + u(\bar{\eta})Z_\xi = -Zf_\epsilon(\Theta) + \dfrac{1}{\mathcal{L}}\Delta Z \ , \end{cases} \tag{2.31}$$

où Δ est l'opérateur de Laplace dans $\mathbb{R}^3$.

Nous nous intéresserons ci-dessous aux solutions d'onde progressive $\Theta(x + ct, \bar{\eta})$, $Z(x + ct, \bar{\eta})$ de (2.31). Remplaçant f_ϵ par f_ϵ^0 pour éviter la difficulté de la frontière froide, nous obtenons le système:

$$\begin{cases} [c + u(\bar{\eta})]\Theta_\xi = Zf_\epsilon^0(\Theta) + \Delta\Theta \ , \\ [c + u(\bar{\eta})]Z_\xi = -Zf_\epsilon^0(\Theta) + \dfrac{1}{\mathcal{L}}\Delta Z \quad \text{dans } S = \mathbb{R} \times \mathcal{O} \ , \end{cases} \tag{2.32}$$

avec les conditions aux limites:

$$\begin{cases} \Theta(-\infty, \bar{\eta}) = 0 \ , \quad Z(-\infty, \bar{\eta}) = 1 \ , \\ \Theta(+\infty, \bar{\eta}) = 1 \ , \quad Z(+\infty, \bar{\eta}) = 0 \quad \text{pour } \bar{\eta} \in \mathcal{O} \ , \end{cases} \tag{2.33}$$

aux extrémités du tube, et:

$$\frac{\partial\Theta}{\partial n} = 0 \ , \quad \frac{\partial Z}{\partial n} = 0 \quad \text{sur } \mathbb{R} \times \partial\mathcal{O} \ , \tag{2.34}$$

sur la paroi du tube (ces dernières conditions expriment que la paroi est adiabatique et imperméable, non catalytique); on a noté n la normale unitaire extérieure sur la paroi $\mathbb{R} \times \partial\mathcal{O}$ du tube S.

Le problème (2.32)-(2.34) est donc l'analogue multi-dimensionnel de (2.28)-(2.29). Les inconnues sont $\Theta(\xi, \bar{\eta})$, $Z(\xi, \bar{\eta})$ et c; le champ de vitesse $u(\bar{\eta})$ est donné.

Lorsque $\mathcal{L} = 1$, il est facile de vérifier que toute solution de (2.32)-(2.34) satisfait encore l'identité $\Theta + Z \equiv 1$. Le problème se réduit de nouveau à une seule équation aux dérivées partielles: il faut trouver Θ et c satisfaisant:

$$[c + u(\bar{\eta})]\Theta_\xi = g_\epsilon^0(\Theta) + \Delta\Theta \quad \text{dans } S = \mathbb{R} \times \mathcal{O} \ , \tag{2.35}$$

$$\Theta(-\infty, \bar{\eta}) = 0 \ , \quad \Theta(+\infty, \bar{\eta}) = 1 \quad \text{pour } \bar{\eta} \in \mathcal{O} \ , \tag{2.36}$$

$$\frac{\partial\Theta}{\partial n} = 0 \quad \text{sur } \mathbb{R} \times \partial\mathcal{O} \ . \tag{2.37}$$

Le problème (2.35)-(2.37) est bien sûr l'analogue multi-dimensionnel de (2.30).

2.4. Flammes non adiabatiques

Les derniers modèles dont nous aurons besoin par la suite concernent la propagation de flammes prémélangées en présence de phénomènes de pertes de chaleur.

Dans le premier de ces modèles, étudié par exemple dans Joulin-Clavin [46], on suppose que la flamme reste plane et que les pertes thermiques apparaissent en volume: recherchant des flammes stationnaires, on considère le problème:

$$\begin{cases} c\Theta_x = Zf_\epsilon^0(\Theta) + \Theta_{xx} - K\Theta \ , \\ cZ_x = -Zf_\epsilon^0(\Theta) + \dfrac{1}{\mathcal{L}} Z_{xx} \ , \end{cases} \tag{2.38}$$

où le flux massique c est encore inconnu, et où $K > 0$ est un coefficient de pertes thermiques volumiques normalisé, avec à la place de (2.29) les conditions aux limites:

$$\begin{cases} \Theta(-\infty) = 0 \ , \quad Z(-\infty) = 1 \ , \\ \Theta(+\infty) = 0 \ , \quad Z_x(+\infty) = 0 \ . \end{cases} \tag{2.39}$$

Dans (2.38), le terme de pertes thermiques $-K\Theta$ est proportionnel à la différence entre la température Θ du mélange et la température des gaz frais (0 en variables normalisées; avec la normalisation de (2.24)-(2.25), ce terme serait proportionnel à $T - T_u^0$).

Nous considérerons aussi l'effet de pertes de chaleur dans le cadre du modèle thermo-diffusif multi-dimensionnel. Suivant Benkhaldoun *et al.* [7], nous supposerons ici que les parois du tube S ne sont pas adiabatiques, et nous considérerons les solutions instationnaires du système:

$$\begin{cases} \Theta_\tau + u(\bar{\eta})\Theta_\xi = Zf_\epsilon(\Theta) + \Delta\Theta \ , \\ Z_\tau + u(\bar{\eta})Z_\xi = -Zf_\epsilon(\Theta) + \dfrac{1}{\mathcal{L}}\Delta Z \quad \text{dans } S = I\!R \times \mathcal{O} \ , \end{cases} \tag{2.40}$$

avec les conditions aux limites:

$$\begin{cases} \Theta(-\infty,\bar{\eta}) = 0 \ , \quad Z(-\infty,\bar{\eta}) = 1 \ , \\ \Theta(+\infty,\bar{\eta}) = 0 \ , \quad Z_\xi(+\infty,\bar{\eta}) = 0 \quad \text{pour } \bar{\eta} \in \mathcal{O} \ , \end{cases} \tag{2.41}$$

$$\frac{\partial\Theta}{\partial n} = -\mathcal{K}\Theta \ , \quad \frac{\partial Z}{\partial n} = 0 \quad \text{sur } I\!R \times \partial\mathcal{O} \ , \tag{2.42}$$

sur la paroi du tube. Ici, $\mathcal{K}$ est un coefficient de pertes thermiques normalisé, et la condition de Neumann pour la température représente une paroi fine faiblement conductive dont la surface externe est maintenue à la tenpérature des gaz frais (0 en variables normalisées). Par rapport au problème (2.32)-(2.34), seules cette condition de paroi sur la température et les conditions aux extrémités du tube ont donc changé.

3. LES LECONS DES PHYSICIENS

Les études des physiciens en théorie de la combustion, et notamment, pour le domaine de la propagation des flammes qui nous intéresse ici, les études asymptotiques pour les grandes valeurs de l'énergie d'activation $\mathcal{E}$, sont plus anciennes que les études numériques et mathématiques dans ce même domaine. Aussi, l'impact de ces études asymptotiques sur les développements numériques et mathématiques est-il grand et bien connu: la plupart des travaux numériques et mathématiques en théorie de la propagation des flammes sont directement inspirés d'analyses réalisées précédemment par les physiciens, les illustrant ou les complétant. Sans chercher à analyser en détail ce rôle des études physiques pour les études numériques et mathématiques, nous allons toutefois en donner quelques exemples.

3.1. Pour les numériciens

3.1.1. Des modèles

Pour les études numériques, le premier apport des études asymptotiques en théorie de la combustion concerne les modèles utilisés. De nombreux modèles simplifiés ont été proposés, qui permettent par exemple d'isoler, dans le phénomène relativement complexe que constitue un écoulement réactif, certains des effets présents en les découplant d'autres effets, au moyen d'une approximation bien choisie, reposant sur des arguments physiques. C'est le cas des modèles isobares et thermodiffusif présentés plus haut; c'est encore le cas d'autres modèles asymptotiques, comme l'équation de Sivashinsky pour la représentation d'un front de flamme plissé [82], [83], ou encore des modèles parfois dits phénoménologiques proposés à la suite des travaux de Darrieus [31], Landau [53] et Markstein [67], où le front de flamme est représenté comme une surface se déplaçant dans un écoulement gazeux avec une vitesse normale relative constante ou dépendant de variables locales comme la courbure du front. Tous ces modèles ont été à de très nombreuses reprises utilisés pour des expériences numériques sur la propagation des flammes (voir par exemple Aimar [3], Ghoniem [39], Majda-Sethian [64], Michelson-Sivashinsky [68] et les références données plus loin).

3.1.2. Des solutions

En plus de ces modèles, les études asymptotiques fournissent des solutions, solutions exactes de modèles simplifiés ou solutions approchées de modèles plus complexes. L'usage le plus fréquent de ces solutions approchées (connues sous forme analytique) par les numériciens consiste à les employer comme donnée initiale pour un algorithme itératif (éventuellement instationnaire) visant à calculer numériquement la solution exacte. L'exemple le plus connu concerne la solution asymptotique (obtenue pour les grandes énergies d'activation) des équations des flammes planes stationnaires (2.28)-(2.29), que nous rappelons ci-dessous (voir par exemple Peters [72]). Plusieurs autres exemples similaires existent dans la littérature, comme celui qui concerne l'utilisation de la solution asymptotique de Burke-Schumann pour le calcul de flammes de diffusion (voir Smooke *et al.* [85]).

Rappelons donc, puisqu'il nous sera utile par la suite, le résultat des analyses asymptotiques pour les grandes énergies d'activation. Pour l'exposer, reprenons la variable $\Theta = \dfrac{T - T_u^0}{T_b^0 - T_u^0}$, et remarquons que le terme d'Arrhenius s'écrit:

$$\exp\left(-\frac{\mathcal{E}}{RT}\right) = \exp\left(-\frac{\mathcal{E}}{RT_b^0}\right) . \exp\left(\frac{-\beta(1-\Theta)}{1-\alpha(1-\Theta)}\right) , \tag{3.1}$$

où $\beta = \dfrac{\mathcal{E}}{RT_b^0}\dfrac{T_b^0 - T_u^0}{T_b^0} > 0$ est l'énergie d'activation réduite (β est aussi appelé le nombre de Zeldovich, c'est l'inverse du paramètre ϵ utilisé plus haut), et $\alpha = \dfrac{T_b^0 - T_u^0}{T_b^0} > 0$ est un paramètre d'élévation de température adimensionné. En supposant pour simplifier que la dépendance en température du taux de réaction ne fait apparaître que le terme exponentiel d'Arrhenius (c'est-à-dire que $r = 0$ dans (2.4)), on peut alors écrire le problème (2.30) sous la forme:

$$\begin{cases} -\Theta'' + c\Theta' = K(1-\Theta)\exp\left(\dfrac{-\beta(1-\Theta)}{1-\alpha(1-\Theta)}\right) = G_\beta(\Theta) \ \text{ sur } \mathbb{R} , \\ \Theta(-\infty) = 0 , \quad \Theta(+\infty) = 1 . \end{cases} \tag{3.2}$$

Appelons (c, Θ) la solution de (3.2). On montre alors, soit en utilisant des arguments empiriques, soit à l'aide de développements asymptotiques raccordés (voir par exemple Buckmaster-Ludford [24], Clavin [28], Clavin-Liñan [29], Larrouturou [58]) que, lorsque β tend vers $+\infty$ (i.e. pour les grandes énergies d'activation), c tend vers:

$$\hat{c} = \lim_{\beta \to +\infty} \sqrt{2 \int_0^1 G_\beta(\Theta) d\Theta} \ , \tag{3.3}$$

pourvu que le second membre de (3.3) soit fini. En particulier, on montre que la solution (c, Θ) de:

$$\begin{cases} -\Theta'' + c\Theta' = \dfrac{\beta^2}{2}(1-\Theta)\exp\left(\dfrac{-\beta(1-\Theta)}{1-\alpha(1-\Theta)}\right) & \text{sur } \mathbb{R} \ , \\ \Theta(-\infty) = 0 \ , \quad \Theta(+\infty) = 1 \ , \end{cases} \tag{3.4}$$

tend lorsque $\beta \to +\infty$ vers (c_0, Θ_0) donné par $c_0 = 1$ et:

$$\Theta_0(x) = \begin{cases} \exp x & \text{si } x \leq 0 \ , \\ 1 & \text{si } x \geq 0 \ . \end{cases} \tag{3.5}$$

Ce résultat fondamental de la théorie de la combustion, qui a été obtenu dès 1938 par Zeldovich et Frank-Kamenetskii [91], donne la structure fondamentale de la flamme prémélangée. Le profil exponentiel de température correspond aux gaz frais et à la zone de préchauffage, et la zone où $\Theta_0 \equiv 1$ est celle des gaz brûlés. De plus, puisque c_0 et Θ_0 vérifient:

$$\begin{cases} -\Theta_0'' + c_0\Theta_0' = c_0\delta \ , \\ \Theta_0(-\infty) = 0 \ , \quad \Theta_0(+\infty) = 1 \ , \end{cases} \tag{3.6}$$

où δ est la masse de Dirac située à l'origine, ce résultat indique aussi que, dans la limite des grandes énergies d'activation, le terme de réaction se comporte comme une masse de Dirac.

Remarque 3: Dans l'état actuel de la formulation du problème, la solution (c, Θ) de (3.4) est définie à une translation de l'origine près. Il faudrait donc donner un sens plus précis à la convergence $(c, \Theta) \to (c_0, \Theta_0)$; nous le ferons dans la Section 3.2.1 ci-dessous. •

Remarque 4: Un résultat analogue existe aussi dans le cas où le nombre de Lewis est différent de 1. Avec des arguments semblables, on montre que la solution

(c, Θ, Z) de:

$$\begin{cases} -\Theta'' + c\Theta' = \dfrac{\beta^2}{2Le} Z \exp\left(\dfrac{-\beta(1-\Theta)}{1-\alpha(1-\Theta)}\right) , \\ -\dfrac{1}{Le} Z'' + cZ' = -\dfrac{\beta^2}{2Le} Z \exp\left(\dfrac{-\beta(1-\Theta)}{1-\alpha(1-\Theta)}\right) \quad \text{sur } I\!R , \\ \Theta(-\infty) = 0 , \quad \Theta(+\infty) = 1 , \\ Z(-\infty) = 1 , \quad Z(+\infty) = 0 , \end{cases} \tag{3.7}$$

tend lorsque $\beta \to +\infty$ vers (c_0, Θ_0, Z_0) donné par $c_0 = 1$ et:

$$\begin{cases} -\Theta_0'' + c_0\Theta_0' = c_0\delta , \\ -\dfrac{1}{\mathcal{L}} Z_0'' + c_0 Z_0' = -c_0\delta , \\ \Theta_0(-\infty) = 0 , \quad \Theta_0(+\infty) = 1 . \\ Z_0(-\infty) = 1 , \quad Z_0(+\infty) = 0 , \end{cases} \tag{3.8}$$

c'est-à-dire:

$$\Theta_0(x) = \begin{cases} \exp x & \text{si } x \leq 0 , \\ 1 & \text{si } x \geq 0 , \end{cases} , \quad Z_0(x) = \begin{cases} 1 - \exp(\mathcal{L}x) & \text{si } x \leq 0 , \\ 0 & \text{si } x \geq 0 . \end{cases} \bullet \tag{3.9}$$

3.1.3. Des informations qualitatives

Enfin, le troisième apport essentiel des études physiques aux numériciens de la combustion concerne toutes les informations qualitatives sur les solutions de nombreux problèmes de la théorie de la combustion.

Un premier exemple de telles informations concerne les *échelles caractéristiques* intervenant dans les phénomènes. Par exemple, outre la forme analytique des solutions asymptotiques, le résultat cité ci-dessus fournit des indications qualitatives importantes: avec la normalisation adoptée, et dans la limite des grandes énergies d'activation, la vitesse de la flamme (donnée par c) vaut 1, son épaisseur (donnée par l'échelle caractéristique de l'exponentielle dans (3.5)) vaut aussi 1. On déduit de ceci que le temps caractéristique de la flamme, appelé temps de transit (rapport de son épaisseur sur sa vitesse) vaut encore 1; on montre aussi que l'épaisseur de la zone de réaction tend vers 0 comme $\dfrac{2}{\beta}$. Bien sûr, on déduit de tous ces résultats, donnés ici pour une normalisation particulière, des résultats analogues pour une normalisation quelconque, et ces informations sont de la première importance pour le numéricien, notamment pour analyser les problèmes de finesse de maillage et de choix du pas de temps (voir par exemple Benkhaldoun-Larrouturou [9], Larrouturou [58]).

D'autres informations qualitatives importantes fournies par les études physiques aux numériciens concernent les instabilités de flamme. C'est le cas en particulier des informations fournies par les analyses asymptotiques sur les seuils d'instabilité, sans lesquelles on peut penser que la plupart des calculs d'instabilités de flamme réalisés à ce jour n'auraient pas été effectués (nous verrons plus loin que ces informations des physiciens sont maintenant complétées par des résultats d'études numériques).

Donnons deux exemples de tels calculs, pour lesquels les résultats des études asymptotiques sont un guide précieux. Nous considérons ici les instabilités cellulaires dites thermo-diffusives, qui résultent uniquement du couplage entre les effets de réaction et de diffusion. Dans le cadre du modèle thermo-diffusif (2.31), ces instabilités ont fait l'objet de nombreuses études, linéaires et non linéaires (voir par exemple [58], [82], [83]; c'est dans ce contexte qu'apparaît en théorie de la combustion l'équation de Sivashinsky). La principale conclusion de l'analyse de stabilité linéaire est que, lorsque l'inégalité $\mathcal{L} \leq \mathcal{L}^{cr} = 1 - \dfrac{2}{\beta}$ a lieu, la flamme est instable pour les perturbations de longueur d'onde supérieure à:

$$\lambda_0 = \frac{2\pi\beta(1-\mathcal{L})}{\sqrt{\frac{\beta}{2}(1-\mathcal{L})-1}} . \tag{3.10}$$

C'est grâce à ce résultat que cette instabilité cellulaire a pu être calculée (voir Guillard *et al.* [42], Bayliss *et al.* [6]). La Figure 1 montre un front de flamme cellulaire stationnaire dans un tube dans la largeur est égale à $L = 3\lambda_0$: pour ce calcul, tiré de [42], on a considéré le modèle thermo-diffusif (2.31), dans un tube de largeur L ($\mathcal{O}$ est réduit à l'intervalle $[0, L]$), avec un champ de vitesse u constant; on a utilisé la normalisation (3.7) (i.e. l'expression du terme de réaction de (3.7)), et les paramètres $\beta = 10$, $\mathcal{L} = 0.7$, $\alpha = 0.8$, $L = 18\sqrt{2}\pi$ (en fait, pour ce calcul, les conditions aux limites de Neumann (2.34) ont été remplacées par des conditions de périodicité, qui permettent d'employer une méthode pseudo-spectrale, plus efficace pour ce problème; des solutions cellulaires analogues ont depuis été calculées avec les conditions de Neumann (2.34) en utilisant une méthode de type éléments finis; voir Benkhaldoun-Larrouturou [9], Dervieux *et al.* [33], [58]). Sur la figure, la taille caractéristique des cellules est bien égale au tiers de la taille L du tube, comme attendu. Nous reviendrons plus loin sur ce résultat (voir la Section 3.2.3).

Figure 1: Front de flamme cellulaire stationnaire (de [42]).

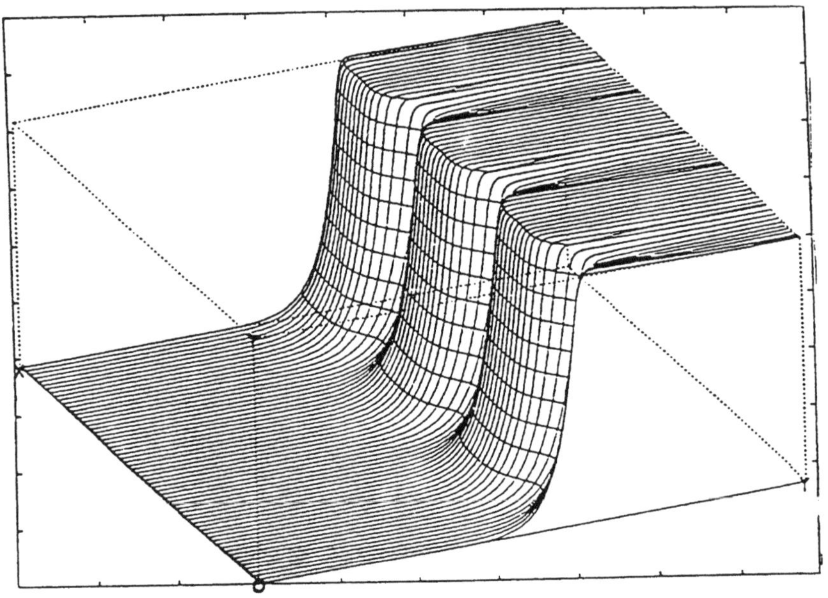

Plusieurs études portant sur l'équation de Sivashinsky suggèrent que, lorsque l'on augmente la taille L du tube, la solution cellulaire stationnaire perd à son tour sa stabilité, et même qu'un régime chaotique peut s'installer, pour L suffisamment grand (voir par exemple Aimar [3]). Ce régime a été observé numériquement sur le modèle thermo-diffusif par Denet [32] (c'est-à-dire en résolvant exactement le même problème que pour la figure précédente, mais avec $\mathcal{L} = 0.6$ et $L = 120$). Les Figures 2 à 4 montrent ce front de flamme thermo-diffusif auto-turbulent, sur lequel on observe les apparitions et disparitions successives de cellules.

Figure 2: Front de flamme cellulaire instationnaire à différents instants successifs $t_1 = 72$, $t_2 = 80$ et $t_3 = 88$ (de [32]).

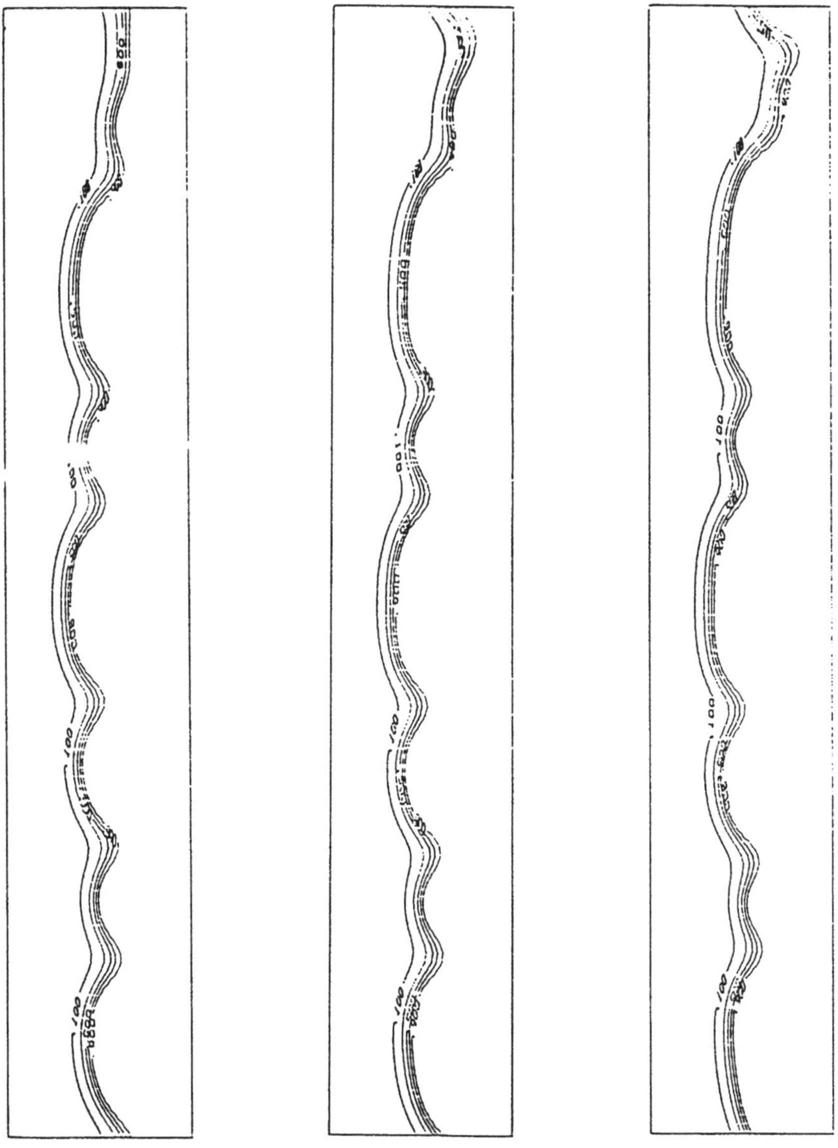

Figure 3: Front de flamme cellulaire instationnaire à différents instants successifs $t_4 = 96$, $t_5 = 104$ et $t_6 = 112$ (de [32]).

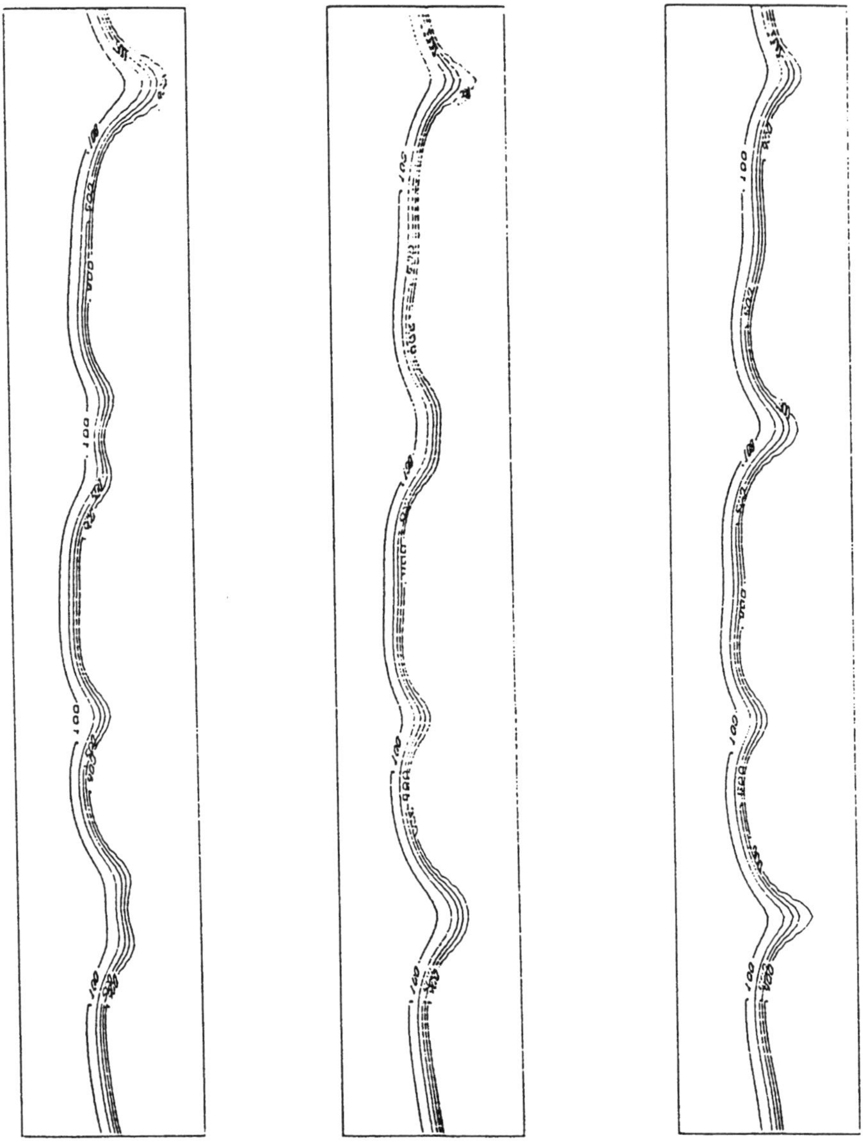

Figure 4: Front de flamme cellulaire instationnaire à différents instants successifs $t_7 = 120$, $t_8 = 128$ (de [32]).

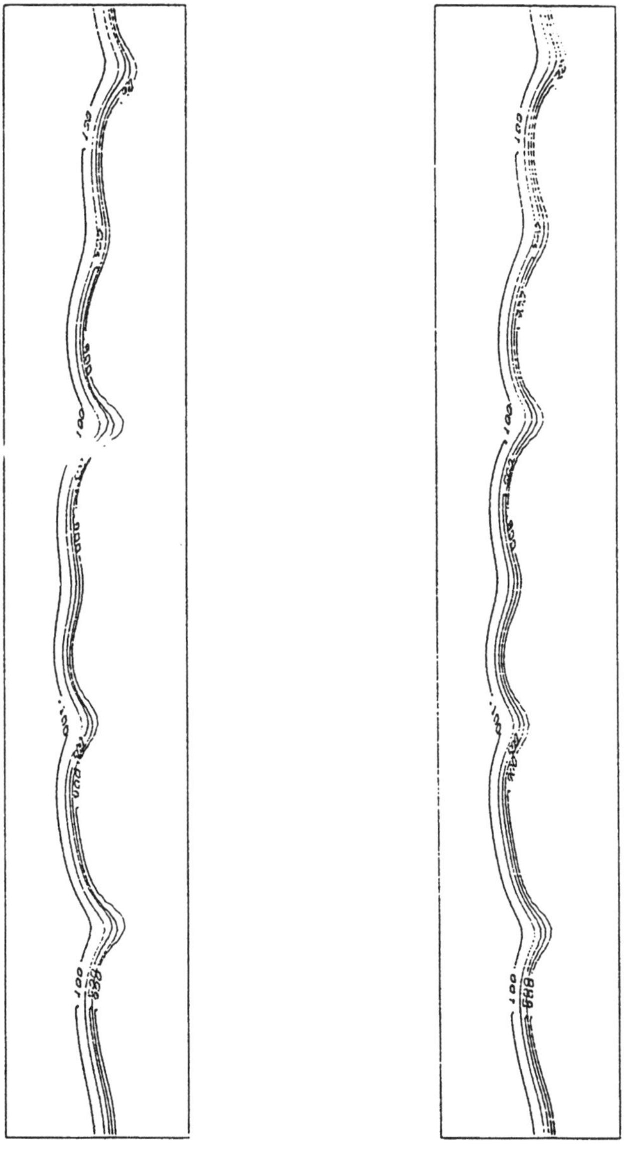

3.2. Pour les mathématiciens

De même que pour les études numériques, les études physiques fournissent les modèles simplifiés et constituent un guide pour les études mathématiques en théorie de la combustion. Ainsi, les premières études mathématiquement rigoureuses ont concerné les solutions d'onde progressive plane (solutions des problèmes (2.30) ou (2.28)-(2.29)), dont l'existence et quelques propriétés qualitatives avaient auparavant été l'objet des études formelles des physiciens (voir les travaux cités dans la Section 2.2 ci-dessus).

Ce guide que fournissent les résultats des physiciens se retrouve dans presque toutes les études mathématiques des modèles de propagation de flammes. Ce guide a d'ailleurs été fort précieux, d'autant plus que les études mathématiquement rigoureuses de ces modèles fortement non linéaires sont fort complexes: pour de nombreux problèmes, l'analyse physique suggére au mathématicien de la combustion les propriétés qu'il peut espérer démontrer, ou au contraire qu'il ne peut espérer prouver. Avant de donner ci-dessous quelques exemples où cette aide des études physiques a été particulièrement nette, mentionnons tout de même ici qu'il existe maintenant quelques exemples opposés où les études mathématiques ont éclairé des problèmes non encore étudiés ou incomplètement étudiés par les physiciens (voir la Section 5.1 ci-dessous).

3.2.1. Analyse asymptotique pour les grandes énergies d'activation

Notre premier exemple concerne les analyses asymptotiques pour les grandes énergies d'activation, dont il faut avouer qu'elles nécessitent souvent une démonstration plus longue et plus complexe avec les méthodes des mathématiciens qu'avec les méthodes formelles des physiciens. Pour ces études mathématiques, nous renvoyons à Berestycki *et al.* [20] et Marion [65], [66] pour des analyses asymptotiques concernant une flamme plane, et à Berestycki *et al.* [10], [22] pour une analyse asymptotique d'une flamme plissée, conduisant à un problème de frontière libre (voir également [12] où plusieurs de ces résultats sont repris).

Nous considérerons ici l'analyse asymptotique de la flamme plane, déjà envisagée dans la Section 3.1.2 ci-dessus. Le résultat formel rappelé dans la Section 3.1.2 fournit ce que les physiciens appellent une "limite distinguée": dans (3.2), il faut étudier la limite des solutions lorsque β tend vers $+\infty$ *tout en supposant* que le préfacteur d'Arrhenius K est proportionnel à β^2.

Ceci est l'information cruciale sur laquelle est basée la démonstration du résultat suivant tiré de Berestycki *et al.* [20]. Ce résultat était jusqu'à très récemment le seul résultat rigoureux concernant la limite des grandes énergies d'activation pour un nombre de Lewis différent de 1. Pour l'exposer, reprenons le système (2.28)-(2.29), et supposons que f_ϵ^0 vérifie (2.27), que:

$$f_\epsilon^0 \text{ est Lipschitzienne sur } [0,1] \ , \tag{3.11}$$

et que la dépendance en ϵ est de la forme:

$$f_\epsilon^0(s) = \frac{1}{\epsilon^2}\Phi\left(\frac{s-1}{\epsilon}\right) \quad \text{pour } \theta \leq s \leq 1 \ , \tag{3.12}$$

où, par exemple, Φ est l'exponentielle $\Phi(\sigma) = e^\sigma$, comme c'est approximativement le cas dans (3.4)); rappelons que $\dfrac{1}{\epsilon^2} = \beta^2$. On suppose de plus que:

$$\Phi \in \ \mathcal{C}^1(I\!R_-, I\!R_+^*) \ , \tag{3.13}$$

$$\exists \ \kappa \in \left]\frac{1}{2}, 1\right[\ , \ \ \lim_{\epsilon \to 0}\left[\sup_{0\leq s\leq \theta_\epsilon} f_\epsilon^0(s)\right] = 0 \ , \quad \text{avec } \theta_\epsilon = 1 - \epsilon^\kappa \ , \tag{3.14}$$

$$\int_{-\infty}^0 |\sigma|\Phi(\sigma)d\sigma = m \quad \text{avec } 0 < m < +\infty \ , \tag{3.15}$$

et (condition purement technique):

$$\Phi \text{ admet un nombre fini d'extrema sur }]-\infty, 0[\ . \tag{3.16}$$

Les conditions (3.12)-(3.15) reviennent à dire que $(1-s)f_\epsilon^0(s)$ tend vers une masse de Dirac "$m\delta_{1-0}(s)$" concentrée à gauche de 1, impliquant en particulier que:

$$\lim_{\epsilon \to 0} \int_0^1 f_\epsilon^0(s)(1-s)ds = m \ . \tag{3.17}$$

Les résultats d'existence et d'analyse asymptotique de [20] sont alors les suivants (voir [12] ou [20] pour la démonstration):

Théorème 1: [20]

Sous les hypothèses (2.27) et (3.11), le problème (2.28)-(2.29) admet au moins une solution $(c_\epsilon, \Theta_\epsilon, Z_\epsilon)$. Si $\mathcal{L} \geq 1$, cette solution est unique à une translation de l'origine près. •

Théorème 2: [20]

On suppose que f_ϵ^0 vérifie (2.27) et (3.11)-(3.16). Soit $(c_\epsilon, \Theta_\epsilon, Z_\epsilon)$ une solution de (2.28)-(2.29) satisfaisant la condition de normalisation:

$$\Theta_\epsilon(0) = \theta \ . \tag{3.18}$$

Lorsque $\epsilon \to 0$, $(c_\epsilon, \Theta_\epsilon, Z_\epsilon)$ converge vers (c_0, Θ_0, Z_0) en ce sens que $c_\epsilon \to c_0$ dans $\mathbb{R}$ et $\Theta_\epsilon \to \Theta_0$, $Z_\epsilon \to Z_0$ dans $\mathcal{C}(\mathbb{R})$. La limite (c_0, Θ_0, Z_0) est donnée par:

$$c_0 = \sqrt{2m\mathcal{L}} \ , \tag{3.19}$$

$$\Theta_0(x) = \begin{cases} \theta \exp(c_0 x) & \text{pour } x \leq \tilde{x} = -\dfrac{1}{c_0}\mathrm{Log}(\theta) \ , \\ 1 & \text{pour } x \geq \tilde{x} \ , \end{cases} \tag{3.20}$$

$$Z_0(x) = \begin{cases} 1 - \exp(c_0 \mathcal{L}(x - \tilde{x})) & \text{pour } x \leq \tilde{x} \ , \\ 0 & \text{pour } x \geq \tilde{x} \ . \bullet \end{cases} \tag{3.21}$$

Une amélioration récente de ce résultat permet maintenant de préciser si la convergence (3.19)-(3.21) reste valable lorsque on fait seulement l'hypothèse (3.17), et non plus l'hypothèse plus forte (3.12) (voir Glangetas [41]).

Remarque 5: La condition (3.18) est utilisée pour briser l'invariance par translation du problème (2.28)-(2.29) (voir Remarque 3 ci-dessus). •

Remarque 6: La limite (3.19) est à rapprocher de (3.3) et (3.17). •

Remarque 7: Le résultat d'existence et d'unicité du Théorème 1 est propre au modèle avec température d'ignition. Si, pour éviter la difficulté de la frontière froide, on introduit une fonction $\hat{f}_\epsilon^0$ vérifiant $\hat{f}_\epsilon^0(0) = 0$ et $\hat{f}_\epsilon^0(s) > 0$ pour $s > 0$ au lieu de (2.27), on obtient un modèle ayant une infinité de solutions (c, Θ, Z) (voir Clavin-Liñan [29], Kolmogorov *et al.* [52], Marion [65], [66]). •

3.2.2. Flamme plane non adiabatique

Le deuxième exemple concerne l'étude du problème (2.38)-(2.39) décrivant une flamme plane en présence de pertes de chaleur volumiques. Ce problème a fait l'objet d'une étude formelle pour les grandes énergies d'activation (voir Joulin-Clavin [46]), en utilisant encore l'expression (3.4) du taux de réaction. Les résultats principaux de cette analyse asymptotique sont les suivants:

Dans la limite des grandes énergies d'activation, la vitesse de la flamme non adiabatique c est donnée par la relation implicite:

$$c^2 \operatorname{Log} c + \beta K = 0 \,. \tag{3.22}$$

Comme le montre la Figure 5, cette équation décrit le processus d'extinction. Il n'y a plus de propagation de flamme dès que le paramètre de pertes thermiques est trop large, c'est-à-dire si:

$$K > K_{ext} = \frac{e^{-1}}{2\beta} \,, \tag{3.23}$$

et la valeur correspondante de la vitesse de flamme est:

$$c_{ext} = \exp\left(-\frac{1}{2}\right) \,. \tag{3.24}$$

Figure 5: Diagramme d'extinction dans le cas de pertes volumiques: $c^2\operatorname{Log}c + \beta K = 0$.

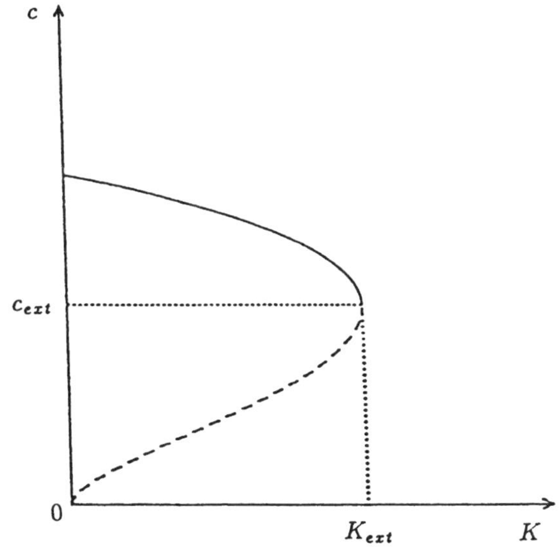

Pour un taux de pertes thermiques inférieur à K_{ext}, deux solutions existent, mais l'analyse montre que seule la plus rapide est stable vis à vis de preturbations planes. Quand K tend vers 0, cette solution plane tend vers la solution adiabatique (3.5), qui se propage à la vitesse $c_0 = 1$.

L'étude mathématique de ce problème a été réalisée par Giovangigli [40]. Cette étude repose de façon essentielle sur les résultats de l'étude asymptotique des physiciens rappelés ci-dessus. Notons $c_{ad} > 0$ la valeur de c correspondant à la solution obtenue pour $K = 0$, solution adiabatique dont l'existence découle du Théorème 1 ci-dessus (nous savons que, avec l'expression (3.4) du taux de réaction, c_{ad} tend vers $c_0 = 1$ lorsque ϵ tend vers 0). Les résultats précédents et la Figure 5 suggèrent que la structure des solutions est plus simple si, au lieu de chercher, comme c'est naturel, la vitesse de la flamme c pour $K > 0$ fixé (inférieur à une valeur limite que l'on ne connaît d'ailleurs que dans la limite des grandes énergies d'activation), on cherche K en fonction de c, pour $0 < c < c_{ad}$. Cette remarque s'est révélée absolument cruciale pour le succés de l'analyse mathématique de [40], dont le résultat principal (outre une analyse asymptotique rigoureuse pour les grandes énergies d'activation qui retrouve les résultats de l'étude formelle) est le suivant:

Théorème 3: [40]

On suppose que f_ϵ^0 vérifie (2.27) et est de classe $\mathcal{C}^1$ et croissante sur $[0,1]$. On suppose aussi que le nombre de Lewis $\mathcal{L}$ est égal à 1, et on appelle $(c_{ad}, \Theta_{ad}, Z_{ad})$ la solution correspondante de (2.28)-(2.29). Alors, pour tout $c \in]0, c_{ad}[$, il existe $K = K(c) > 0$ et Θ, $Z \in \mathcal{C}^2(\mathbb{R})$ solution de (2.38)-(2.39). De plus, on a:

$$\lim_{c \searrow 0} K(c) = 0 \ , \quad \lim_{c \nearrow c_{ad}} K(c) = 0 \ . \bullet \tag{3.25}$$

Remarque 8: Le résultat (3.25) montre que la courbe $c \mapsto K(c)$ a bien la même allure que sur la Figure 5. Il faut cependant noter ici que l'unicité de $K(c)$ n'est pas démontrée; les limites (3.25) ont lieu pour tout $K(c)$ tel qu'une solution (Θ, Z) de (2.38)-(2.39) existe. •

3.2.3. Modèle thermo-diffusif multi-dimensionnel

Dans notre troisième exemple, les études formelles des physiciens vont moins apparaître comme une aide suggérant au mathématicien une méthode ou un résultat que comme un guide soulignant les difficultés et indiquant quelles propriétés on ne peut espérer démontrer.

Considérons donc le modèle thermo-diffusif multi-dimensionnel (2.32)-(2.34), qui se résuit à (2.35)-(2.37) lorsque $\mathcal{L} = 1$. Alors que l'étude mathématique du problème (2.32)-(2.34) (cas où $\mathcal{L} \neq 1$) est actuellement à peine entamée (voir

Berestycki *et al.* [16] pour un premier résultat dans cette direction), tandis que plusieurs résultats existent pour le problème (2.35)-(2.37) (cas où $\mathcal{L} = 1$; voir Berestycki *al.* [10], [11], [15], [17], [19], [21], [22] et la Section 5.1.1 ci-dessous), on sait sur la base des études physiques que les solutions de (2.32)-(2.34) et celles de (2.35)-(2.37) présentent quelques propriétés qualitatives différentes (ce point explique d'ailleurs les difficultés supplémentaires rencontrées dans l'étude du système). En particulier, s'il est intuitivement évident d'un point de vue physique, et facile à vérifier rigoureusement en utilisant le principe du maximum, qu'une solution du système (2.32)-(2.34) vérifie les inégalités:

$$0 \le Z \le 1 \ , \quad 0 \le \Theta \ \text{ dans } S \ , \tag{3.26}$$

il n'est plus possible quand $\mathcal{L} \neq 1$ de montrer que cette solution vérifie l'inégalité $\Theta \le 1$, et il semble difficile de trouver (rigoureusement ou de façon heuristique) une estimation a priori du type $\Theta \le \hat{\Theta}$. Les études des physiciens montrent d'ailleurs que la température Θ peut effectivement dépasser 1 lorsque $\mathcal{L} \neq 1$ (voir par exemple Sivashinsky [81]). La solution n'est alors plus monotone croissante dans la direction ξ, alors qu'elle l'est si $\mathcal{L} = 1$: on a dans ce cas $\Theta \le 1$, $\Theta_\xi \ge 0$ (voir la Section 5.1.1).

Une autre différence prévisible (sur la base des études formelles) entre les problèmes (2.32)-(2.34) et (2.35)-(2.37) concerne l'unicité de la solution. Pour faire apparaître cette différence, considérons le cas où le champ de vitesse u est constant dans (2.32) et (2.35). Le résultat d'unicité valable pour le cas $\mathcal{L} = 1$ (voir la Section 5.1.1 ci-dessous) s'interprète alors de la façon suivante:

Théorème 4: [22]

On suppose que l'hypothèse (2.27) est vérifiée, et que $u \equiv 0$. Lorsque $\mathcal{L} = 1$, le système (2.32)-(2.34) a une solution unique $(c, \Theta(\xi, \bar{\eta}), Z(\xi, \bar{\eta}))$ à une translation de l'origine près. Cette solution est plane: $\Theta(\xi, \bar{\eta}) = \Theta(\xi)$, $Z = 1 - \Theta$; c'est la solution de (2.30). •

Lorsque $\mathcal{L} = 1$ et que u est constant, le système (2.32)-(2.34) a donc comme seule solution la solution plane donnée par (2.30). Les arguments formels des physiciens indiquent que cette propriété n'est plus assurée lorsque $\mathcal{L} < 1$. En effet, il existe toujours (au moins) une solution monodimensionnelle lorsque $\mathcal{L} < 1$ d'après le Théorème 1. Mais, pour les grandes valeurs de l'énergie d'activation, cette solution plane est instable vis à vis de perturbations multidimensionnelles (voir par exemple [24], [28], [58], [82]), et une autre solution stationnaire stable réellement multidimensionnelle existe. Le problème (2.32)-(2.34) a alors au moins deux solutions.

Cette perte d'unicité conjecturée sur la base des analyses asymptotiques des physiciens peut d'ailleurs être "vérifiée" numériquement. Lorsque $\mathcal{L} < 1$, on peut à la fois calculer la solution plane (instable) par un calcul monodimensionnel (voir par exemple Larrouturou [57], Peters-Warnatz [74]), et la solution stable plissée par un calcul multidimensionnel (voir Figure 1 ci-dessus). On observe d'ailleurs sur cette solution plissée que les inégalités $\Theta \leq 1$ et $\Theta_\xi \geq 0$ ne sont pas satisfaites.

4. LES LECONS DES NUMERICIENS

Examinons maintenant l'apport des études numériques, en gardant présent à l'esprit que nous nous limitons aux apports touchant aux problèmes "fondamentaux" de la théorie de la combustion, tels que ceux que nous avons déjà évoqués. Certainement, ces enseignements des travaux numériques seraient comparativement bien plus grands si nous considérions des problèmes moins académiques, qui seraient souvent trop complexes pour les études asymptotiques et plus encore pour l'analyse mathématique. C'est d'ailleurs ce qui s'est produit longtemps pour les problèmes de propagation de flammes avec chimie complexe, domaine dans lequel de très nombreux travaux basés sur des expériences numériques (voir par exemple [35], [38], [49], [50], [51], [70], [74], [79], [84], [86], [87]) ont précédé les quelques études asymptotiques qui se sont développées récemment (voir par exemple [73], [75], [80]).

4.1. Pour les physiciens

Outre l'exemple des problèmes de chimie complexe, nous donnerons deux exemples où des travaux numériques ont soit précédé les études physiques, soit apporté des informations sur des questions situées hors de la portée actuelle des analyses formelles.

4.1.1. Instabilités non linéaires

Le premier exemple concerne certaines problèmes d'instabilités de flamme. Certes, les études physiques ont été nombreuses et importantes pour analyser différents phénomènes d'instabilité (voir par exemple Clavin [28], Clavin *et al.*

[30], Darrieus [31], Landau [53], Pelcé-Clavin [71], Sivashinsky [82], [83]), et, nous l'avons dit, ces études ont servi de guides aux premiers travaux numériques sur la question. Mais ces études concernent pour une bonne part des analyses linéaires ou faiblement non linéaires de stabilité, et analysent surtout les seuils d'instabilité. Et il est clair que l'apport des travaux numériques est primordial pour compléter ces études en cherchant à analyser les phénomènes fortement non linéaires, les régimes de bifurcation successives, loin au delà des seuils. C'est par exemple le cas des travaux sur les phénomènes successifs de doublement de période pour les problèmes d'instabilité pulsatoire apparaissant pour les flammes planes instationnaires (système (2.28)-(2.29) aux grandes valeurs du nombre de Lewis; voir Bayliss *et al.* [6]). C'est encore le cas des études numériques allant jusqu'à l'étude d'un régime auto-turbulent pour l'équation de Sivashinsky (voir par exemple Aimar [3], Michelson-Sivashinsky [68]) ou pour le modèle thermo-diffusif (voir Denet [32] et les Figures 2 à 4 ci-dessus).

Il est aussi des types d'instabilités de combustion qui, si elles sont reconnues comme "fondamentales", font intervenir le couplage entre un nombre élevé d'effets physiques élémentaires, ce qui les rend difficiles à étudier analytiquement. Nous pensons par exemple à l'instabilité dite de "flamme tulipe", identifiée depuis Markstein [67], qui se produit lors de la propagation d'une flamme prémélangée dans une boîte rectangulaire fermée et suffisamment allongée. Pour ce type de problèmes, que l'on commence à pouvoir étudier numériquement (voir par exemple N'Konga *et al.* [69], ...), on peut penser que les travaux numériques devraient jouer un rôle d'aide à une meilleure compréhension, au côté des études expérimentales.

4.1.2. Flammes plissées non adiabatiques

Le deuxième exemple concerne l'étude de l'effet de pertes de chaleur pariétales sur une flamme prémélangée, effet que l'on peut d'abord chercher à analyser dans le cadre du modèle thermo-diffusif (2.39)-(2.41). Si nous avons vu que, dans le cadre monodimensionnel des pertes thermiques volumiques, l'analyse asymptotique a précédé et guidé les travaux numériques et mathématiques, à l'opposé pour le problème multi-dimensionnel, ce sont les travaux numériques qui furent développés les premiers (voir Benkhaldoun *et al.* [7]), et qui ont suscité depuis un début d'analyse formelle [62].

Présentons rapidement l'essentiel de ces résultats numériques. Dans les essais numériques de [7], on résoud le problème instationnaire (2.40)-(2.42), avec $\mathcal{L} = 1$, $u \equiv 0$, avec l'expression (3.4) du taux de réaction, et en utilisant comme donnée initiale les profils de température et de fraction massique (3.9) de la flamme plane asymptotique adiabatique. Au cours d'un calcul, on observe que la flamme ralentit, que sa courbure augmente, et qu'une zone d'extinction se propage à partir des parois du tube vers son axe. Comme on l'attend, on obtient deux comportements différents pour les grandes valeurs du temps: si le coefficient de pertes thermiques pariétales $\mathcal{K}$ est suffisamment grand, la température Θ décroît vers 0 sur l'ensemble du domaine et la flamme s'éteint; au contraire, pour des petites valeurs de $\mathcal{K}$ la vitesse instantanée de la flamme converge vers une valeur non nulle, et l'on observe finalement une flamme non adiabatique stationnaire (ou, ce qui est équivalent, se déplaçant à vitesse constante).

Un exemple de ces flammes non adiabatiques stationnaires est présenté sur la Figure 6. On observe que le taux de réaction chimique est très faible au voisinage des parois, à cause du refroidissement de ces régions, et que la température des gaz brûlés décroît derrière la flamme.

Figure 6: Flamme non adiabatique stationnaire, pour $\beta = 10$, $\mathcal{L} = 1$, $\alpha = 0.84$, $L = 16$, $\mathcal{K} = 0.2$: courbes isovaleurs pour la température, la fraction massique et le taux de réaction.

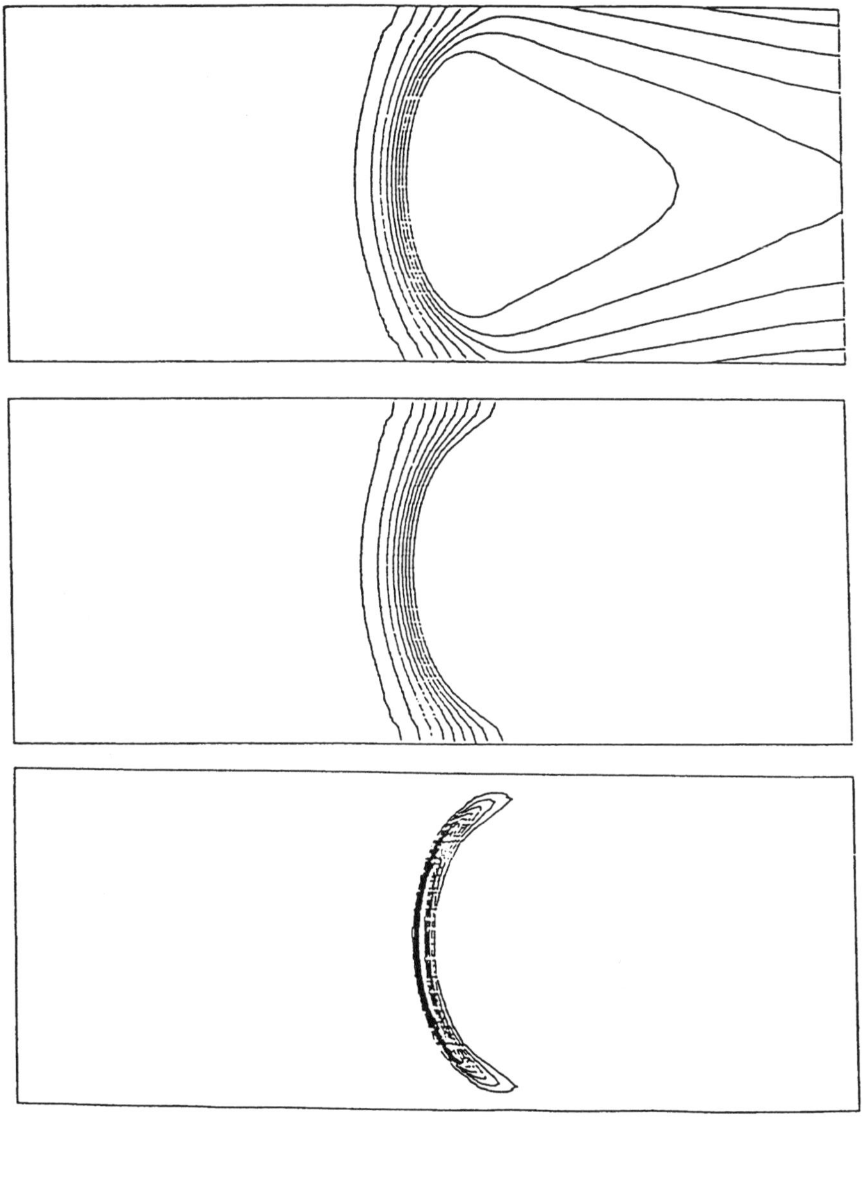

On montre aussi dans [7], sur la base des résultats numériques, que la valeur limite $\mathcal{K}_{ext}$ du coefficient de pertes à partir de laquelle se produit l'extinction est proportionnelle à l'inverse de l'énergie d'activation réduite, comme c'est le cas pour le modèle de pertes volumiques (2.38)-(2.39) (voir (3.23)). Par contre, les résultats des essais numériques multi-dimensionnels diffèrent des conclusions de l'analyse monodimensionnelle rappelés dans la Section 3.2.2 par d'autres aspects, notamment en ce qui concerne la valeur de la vitesse de flamme à la limite d'extinction. Nous renvoyons à [7] pour les détails.

4.2. Pour les mathématiciens

Nous avons déjà vu ci-dessus, par exemple dans la Section 3.2.3, que les travaux numériques pouvaient, au même titre que les études physiques, guider l'analyse mathématique des modèles de la théorie de la combustion. Nous verrons encore dans la Section 5.1.1 ci-dessous un exemple de résultat mathématique concernant le modèle thermo-diffusif multi-dimensionnel qui fut d'abord conjecturé sur la base d'essais numériques et sans le secours de conclusions d'études formelles avant d'être démontré rigoureusement. C'est aussi notre expérience que, dansbien d'autres cas que nous ne développerons pas ici, les essais numériques ont joué un rôle important en accompagnant et en guidant l'analyse mathématique, comme par exemple pour plusieurs des résultats de [36] (voir la Section 5.2.2 ci-dessous).

5. LES LECONS DES MATHEMATICIENS

5.1. Pour les physiciens

Si les études mathématiques dans le domaine de la combustion ont beaucoup bénéficié des études formelles, et ont pendant longtemps surtout apporté des preuves rigoureuses de résultats déjà connus des physiciens, il existe maintenant plusieurs exemples d'études mathématiques dont les résultats sont instructifs pour la communauté des physiciens de la combustion. Nous en donnons ci-dessous deux exemples.

5.1.1. Analyse du modèle thermo-diffusif multi-dimensionnel

Le modèle thermo-diffusif (2.32)-(2.34) (ou sa version plus simple (2.35)-(2.37) lorsque $\mathcal{L} = 1$), dans lequel on a supprimé l'influence de la combustion sur le champ des vitesses, constitue le modèle le plus simple permettant d'étudier l'action d'un champ de vitesse non uniforme $u(\bar{\eta})$ sur une flamme plissée et est à ce titre particulièrement intéressant. La question générale de l'existence d'une solution d'onde progressive (solution de (2.32)-(2.34)) pour tout champ de vitesse u, même fortement non uniforme, est loin d'être tranchée, et on peut se demander si, dans le cadre de ce modèle, un champ de vitesse fortement cisaillé a ou non pour effet d'éteindre la flamme (voir Berestycki-Sivashinsky [23]).

Alors qu'il a peu été étudié par des méthodes formelles (mis à part [23], sur laquelle nous reviendrons plus loin), ce modèle a fait l'objet, dans le cas où $\mathcal{L} = 1$, de plusieurs études mathématiques, dont nous présentons maintenant les principaux résultats.

Le premier résultat important est le suivant:

Théorème 5: [15], [22]

On suppose que:

$$\begin{cases} g_\epsilon^0 \text{ est Lipschitzienne sur } [0,1] \ , \\ \exists \theta \in]0,1[\ , \quad g_\epsilon^0 \equiv 0 \text{ sur } [0,\theta] \ , \quad g_\epsilon^0 > 0 \text{ sur }]\theta,1[\ , \quad g_\epsilon^0(1) = 0 \ , \end{cases} \tag{5.1}$$

et que le champ de vitesse donné $u(\bar{\eta})$ est continu sur $\mathcal{O}$. On note $< u > = \dfrac{1}{|\mathcal{O}|}\displaystyle\int_{\mathcal{O}} u(\bar{\eta})d\bar{\eta}$. Le problème (2.35)-(2.37) admet alors une solution unique (c,Θ), qui satisfait:

$$<u> + c > 0 \ , \tag{5.2}$$

$$0 \leq \Theta(\xi,\bar{\eta}) \leq 1 \quad \text{pour tout } (\xi,\bar{\eta}) \in S \ , \tag{5.3}$$

$$\Theta_\xi > 0 \quad \text{dans } S \ . \bullet \tag{5.4}$$

La preuve de ce résultat est bien sûr nettement plus complexe que celle du résultat analogue dans le cas monodimensionnel (Théorème 1, pour $\mathcal{L} = 1$). Elle utilise notamment des estimations du type estimations de Schauder ou issues du principe du maximum pour des équations elliptiques, un argument de degré topologique, et surtout un résultat de monotonie de solutions d'équations elliptiques semi-linéaires, démontré dans Berestycki-Nirenberg [21]. L'existence de la solution a été démontrée par Berestycki *et al.* [15], et son unicité Berestycki-Nirenberg [22] (voir également [17]).

L'existence de la solution (c, Θ) avait été démontrée d'abord, et avec des arguments mathématiques un peu plus simples, par Berestycki-Larrouturou [11], sous l'hypothèse supplémentaire que le champ de vitesse u "n'est pas trop fortement non uniforme". Plus précisément, on montre dans [11] le résultat suivant:

Proposition 6: [11]

On suppose que les hypothèses du Théorème 5 sont vérifiées, et que:

$$\left(\max_{\bar\eta\in\mathcal{O}} u(\bar\eta) - \min_{\bar\eta\in\mathcal{O}} u(\bar\eta)\right) < \sqrt{2\int_0^1 g_\epsilon^0(s)ds}\ . \tag{5.5}$$

Alors la solution (c, Θ) de (2.35)-(2.37) satisfait:

$$c + \min_{\bar\eta\in\mathcal{O}} u(\bar\eta) > 0\ . \bullet \tag{5.6}$$

Par opposition à cette situation, la solution obtenue pour un champ de vitesse u quelconque ne satisfait pas nécessairement (5.6): $c + u(\bar\eta)$ peut changer de signe. Ce phénomène s'interprète comme une "inversion du champ de vitesse". En effet, (2.35) montre que $c + u(\bar\eta)$ s'interprète comme la vitesse du mélange dans le repère $\mathcal{R}_f$ dans lequel la solution est stationnaire (le repère $\mathcal{R}_f$, attaché à la flamme, se déplace à la vitesse $-c$ par rapport au repère d'origine). Donc, (5.6) montre que, la vitesse du mélange dans le repère $\mathcal{R}_f$ est en tout point dirigée vers la droite, c'est-à-dire des gaz frais vers les gaz brûlés, une situation naturelle d'un point de vue physique. Mais, pour des solutions telles que:

$$c + \min_{\bar\eta\in\mathcal{O}} u(\bar\eta) < 0\ , \tag{5.7}$$

il existe des régions du tube S (là où $c + u(\bar\eta) < 0$) où la vitesse, toujours mesurée dans le repère $\mathcal{R}_f$, est dirigée dans la direction opposée, des gaz brûlés vers le mélange frais ! Il est important de noter ici que de telles solutions, "non classiques", ne sont pas "non physiques": il est connu depuis longtemps (pas pour une flamme dans un tube, mais dans d'autres configurations géométriques; voir par exemple Williams [87, p. 418]) que la vitesse *du mélange gazeux* au voisinage d'une flamme peut être dirigée vers les gaz frais, ce qui signifie simplement que les effets convectifs sont localement dominés par les effets diffusifs. Dans tous les cas, (5.2) indique que la vitesse moyenne dans le repère $\mathcal{R}_f$ est positive.

On prouve de plus dans [15] qu'il existe bien des solutions de (2.35)-(2.37) qui exhibent ce phénomène d'inversion du champ de vitesse (i.e. qui satisfont (5.7)). On montre d'ailleurs que la condition (5.5), qui assure que l'inversion du champ de vitesse ne se produit pas, est optimale au sens suivant:

Proposition 7: [15]

Pour tout réel $\delta > 0$, il existe des fonctions u et g_ϵ^0 vérifiant les hypothèses du Théorème 5 et les inégalités:

$$\sqrt{2\int_0^1 g_\epsilon^0(s)ds} < \left(\max_{\bar\eta\in\mathcal{O}} u(\bar\eta) - \min_{\bar\eta\in\mathcal{O}} u(\bar\eta)\right) < \sqrt{2\int_0^1 g_\epsilon^0(s)ds} + \delta\ , \qquad (5.8)$$

et telles que la solution correspondante (c, Θ) de (2.35)-(2.37) satisfait (5.7). •

La démonstration de ce dernier résultat utilise d'ailleurs le fait que la quantité $\sqrt{2\int_0^1 g_\epsilon^0(s)ds}$ est la vitesse asymptotique de la flamme plane pour de grandes énergies d'activation, comme le montrent (3.3) et (3.19).

Comme nous l'avons dit dans la Section 4.2 ci-dessus, le résultat d'existence d'une solution (c, Θ) de (2.35)-(2.37) pour un champ de vitesse u quelconque a d'abord été conjecturé dans [11] au vu d'essais numériques réalisés par Benkhaldoun et Larrouturou [9]. Dans ces expériences numériques, on considère de nouveau un tube de largeur L ($\mathcal{O} =]0, L[$) et on prescrit un champ de vitesse sinusoïdal: $u(\eta) = A\sin\left(\frac{\pi\eta}{L}\right)$, et on observe que l'inversion du champ de vitesse dans le repère attaché à la flamme stationnaire se produit dès que le champ de vitesse est suffisamment cisaillé, c'est-à-dire dès que l'amplitude A est assez grande.

Ces résultats apportent d'ailleurs une lumière nouvelle sur les conclusions de l'unique analyse formelle réalisée sur le système thermo-diffusif dans le cas d'un champ de vitesse non uniforme par Berestycki-Sivashinsky [23]: dans cette étude [23], on examine le problème (2.32)-(2.34), i.e. le cas d'un nombre de Lewis différent de 1, dans un tube bidimensionnel, avec une hypothèse de périodicité dans la direction η, avec encore un champ de vitesse sinusoïdal $u(\eta) = A\sin\left(\frac{\pi\eta}{L}\right)$, et on étudie le développement asymptotique formel pour les grandes énergies d'activation des solutions (c, Θ, Z) de (2.32)-(2.34) *en supposant que l'inversion du champ de vitesse ne se produit pas* ($c + u(\eta) > 0$ pour tout η). On observe alors que le développement asymptotique de ces solutions n'existe que si l'amplitude A du champ de vitesse n'est pas trop grande, ce que les auteurs interprètent en disant qu'une extinction de la flamme se produit lorsque l'on augmente l'amplitude du cisaillement A. Au vu des résultats mathématiques et numériques précédents (obtenus seulement dans le cas plus simple où $\mathcal{L} = 1$), il semble bien que l'augmentation de l'amplitude A produise non l'extinction de la flamme mais plus simplement le phénomène d'inversion du champ de vitesse. La question de savoir si le résultat d'existence d'une flamme stationnaire pour un champ de vitesse quelconque (Théorème 5) reste vrai pour un nombre de Lewis quelconque est d'ailleurs actuellement totalement ouverte.

Pour terminer ce paragraphe, nous indiquons encore deux résultats mathématiques intéressants du point de vue physique pour le problème (2.35)-(2.37) (cas $\mathcal{L} = 1$).

Signalons d'abord que l'analyse rigoureuse de la stabilité des solutions stationnaires de (2.35)-(2.37) vient d'être effectuée par Berestycki *et al.* [19] et Roquejoffre [77]: quelque soit le champ de vitesse u (i.e. même si le phénomène d'inversion du champ de vitesse se produit), la solution stationnaire de (2.35)-(2.37) est stable vis à vis de perturbations multi-dimensionnelles. On présente dans [19] l'analyse linéaire de stabilité et dans [77] l'analyse non linéaire (en considérant le problème instationnaire avec une donnée initiale proche de la solution stationnaire).

Considérons enfin l'analyse asymptotique du problème (2.35)-(2.37) pour les grandes énergies d'activation. Nous supposerons encore ici que $\mathcal{O}$ est un intervalle ouvert de $\mathbb{R}$: $\mathcal{O} =]0, L[$, et que g_ϵ^0 vérifie (5.1) et les hypothèses suivantes:

$$g_\epsilon^0 \in \mathcal{C}^1([0,1], \mathbb{R}_+) \ , \quad {g_\epsilon^0}'(1) < 0 \ , \tag{5.9}$$

$$\forall\, \delta > 0, \ \lim_{\epsilon \to 0} \left[\max_{0 \le s \le 1-\delta} g_\epsilon^0(s) \right] = 0 \ , \tag{5.10}$$

$$\lim_{\epsilon \to 0} \int_0^1 g_\epsilon^0(s) ds = m > 0 \ . \tag{5.11}$$

On notera dans la suite $(c_\epsilon, \Theta_\epsilon)$ l'unique solution de (2.35)-(2.37) associée au terme de réaction g_ϵ^0 et vérifiant la condition de normalisation:

$$\max_{\eta \in \mathcal{O}} \Theta_\epsilon(0, \eta) = \theta \ . \tag{5.12}$$

Les conditions (5.9)-(5.11) sont plus générales que les hypothèses (3.12)-(3.16) retenues pour l'étude asymptotique du système en dimension 1. Cependant, l'étude de la limite singulière du problème (2.35)-(2.37) lorsque $\epsilon \to 0$ est beaucoup plus délicate en dimension supérieure et conduit en particulier à un problème de frontière libre d'un type nouveau. Nous citons simplement un résultat de Berestycki-Nirenberg [22]:

Théorème 8: [22]

Soit (ϵ_n) une suite de réels positifs tendant vers 0. Sous les hypothèses (5.1) et (5.9)-(5.11), on peut extraire une sous-suite (encore notée ϵ_n) telle que les solutions $(c_{\epsilon_n}, \Theta_{\epsilon_n})$ de (2.35)-(2.37) et (5.12) correspondant à $g^0_{\epsilon_n}$ vérifient les propriétés suivantes:

$$c_{\epsilon_n} \to c \text{ avec } 0 < c < +\infty , \tag{5.13}$$

$$\Theta_{\epsilon_n} \to \Theta \text{ dans } \mathcal{C}(\bar{S}) \cap H^1_{loc}(S) . \tag{5.14}$$

La limite Θ est continue dans $\bar{S}$, croissante (au sens large) dans la direction ξ, et vérifie $\Theta(-\infty, \eta) = 0$ pour $\eta \in \mathcal{O}$, $\dfrac{\partial\Theta}{\partial n} = 0$ sur $\partial S = I\!R \times \partial\mathcal{O}$. De plus:

$$\exists\, \hat{\xi} > 0 , \quad \Theta(\xi, \eta) = 1 \;\; \forall \xi \geq \hat{\xi} , \quad \forall \eta \in \mathcal{O} . \tag{5.15}$$

En notant $\mathcal{S}^- = \{(\xi, \eta) \in S,\ \Theta(\xi, \eta) < 1\}$, on a aussi:

$$-\Delta\Theta + cu(\eta)\Theta_\xi = 0 \text{ dans } \mathcal{S}^- . \tag{5.16}$$

Enfin, si l'on suppose que $\Gamma = \partial\mathcal{S}^- \cap S$ est régulière ($\mathcal{C}^1$ par morceaux), et en notant Θ^- la restriction de Θ sur $\mathcal{S}^-$ et ν la normale unitaire extérieure à $\mathcal{S}^-$ sur Γ, on a:

$$\frac{\partial\Theta^-}{\partial\nu} = \sqrt{2m} \text{ sur } \Gamma . \bullet \tag{5.17}$$

Comme on le voit, la limite (c, Θ) est solution d'un problème de frontière libre (à savoir Γ) qui s'écrit:

$$-\Delta\Theta + cu(\eta)\Theta_\xi = 0 \text{ dans } \mathcal{S}^- , \tag{5.18}$$

$$\Theta(-\infty, \eta) = 0 \text{ pour } \eta \in \mathcal{O} , \tag{5.19}$$

$$\frac{\partial\Theta}{\partial n} = 0 \text{ sur } \partial S \cap \partial\mathcal{S}^- , \tag{5.20}$$

$$\Theta = 1 \text{ et } \frac{\partial\Theta}{\partial\nu} = \sqrt{2m} \text{ sur } \Gamma . \tag{5.21}$$

Les questions de régularité de la frontière libre, d'unicité de la solution (c, Θ) du problème (5.18)-(5.21) et, partant, de la convergence globale des solutions $(c_\epsilon, \Theta_\epsilon)$ vers la limite (c, Θ) sont étudiées par Berestycki *et al.* [10].

5.1.2. Analyse de la difficulté de la frontière froide

Nous décrivons maintenant un traitement mathématique rigoureux de la difficulté de la frontière froide, dont nous avons déjà parlé dans la Section 2.2.2 ci-dessus.

Cette difficulté provient donc de ce que le problème (2.24)-(2.25) est mal posé, car le mélange frais n'est pas dans un état d'équilibre chimique: $Y_u^0 f_\epsilon(T_u^0) \neq 0$. En fait, cette difficulté est essentiellement mathématique: la valeur du taux de réaction dans les gaz frais est certes non nulle, mais elle est extrêmement faible par rapport à la valeur du taux de réaction dans la flamme. Ceci est relié au fait que l'énergie d'activation $\mathcal{E}$ est grande, plus exactement que le rapport $\dfrac{\mathcal{E}}{RT}$ est grand pour des températures variant dans l'intervalle considéré $[T_u^0, T_b^0]$. En pratique, le rapport $\dfrac{e^{-\mathcal{E}/RT_u^0}}{e^{-\mathcal{E}/RT_b^0}}$ peut être de l'ordre de e^{-50} ! En d'autres termes, le temps caractéristique τ_u de la réaction chimique dans le mélange frais est extrêmement grand (un grand nombre de millions d'années !). On peut donc penser (si l'on accepte la modélisation du taux de réaction par la loi d'Arrhenius) que les flammes "stationnaires" qui sont observées expérimentalement ne sont pas réellement stationnaires, mais qu'elles évoluent sur des échelles de temps très grandes, de l'ordre de τ_u. On peut alors penser que, bien que le problème (2.6) n'admette pas, tel quel (sans modification du terme d'Arrhenius), de solution stationnaire, ce problème a une solution instationnaire qui "reste très longtemps très proche d'une flamme stationnaire". C'est ce que montre rigoureusement l'analyse mathématique de Roquejoffre [76] et Roquejoffre it et al. [78] (voir également [18]), dont nous présentons ci-dessous les principaux résultats.

Commençons en décrivant le résultat d'existence de solutions instationnaires. Nous considérons donc le problème (2.17)-(2.20), et introduisons une fonction Γ satisfaisant:

$$\begin{cases} \Gamma \in \mathcal{C}^\infty(\mathbb{R}, \mathbb{R}) \,, \\ \Gamma \equiv 0 \text{ on }]-\infty, -1] \,, \quad \Gamma \equiv 1 \text{ on } [1, +\infty[\,. \end{cases} \tag{5.22}$$

Posant:

$$\begin{cases} T^0(x) = T_u^0 + (T_b^0 - T_u^0)\Gamma(x) + \phi^0(x) \,, \\ Y^0(x) = Y_u^0 + (Y_b^0 - Y_u^0)\Gamma(x) + \psi^0(x) \,, \end{cases} \tag{5.23}$$

nous supposerons que:

$$\phi^0 \in H^4(\mathbb{R}) \cap L^1(\mathbb{R}) \,, \quad \psi^0 \in H^4(\mathbb{R}) \cap L^1(\mathbb{R}) \,, \tag{5.24}$$

$$T_{min} = \inf_{x \in \mathbb{R}} T^0(x) > 0 \,, \quad \inf_{x \in \mathbb{R}} Y^0(x) \geq 0 \,, \tag{5.25}$$

et que:

$$\begin{cases} f_\epsilon \in \mathcal{C}^3(\mathbb{R}_+, \mathbb{R}_+) \cup \mathrm{L}^\infty((\mathbb{R}_+) , \\ f_\epsilon \text{ est Lipschitzienne sur } \mathbb{R}_+ . \end{cases} \tag{5.26}$$

On a alors les deux résultats suivants:

Proposition 9: [76]

Sous les hypothèses (5.24)-(5.26), le système différentiel (2.20) a une solution unique $[T_u(t), T_b(t), Y_u(t), Y_b(t)]$. Cette solution est globale (existe pour tout t) et bornée, et satisfait:

$$\begin{cases} T_u(t) > 0 , \quad T_u'(t) \geq 0 , \\ T_b(t) > 0 , \quad T_b'(t) \geq 0 , \\ Y_u(t) \geq 0 , \quad Y_u'(t) \leq 0 , \quad \lim_{t \to +\infty} Y_u(t) = 0 , \\ Y_b(t) \geq 0 , \quad Y_b'(t) \leq 0 , \quad \lim_{t \to +\infty} Y_b(t) = 0 . \bullet \end{cases} \tag{5,27}$$

Théorème 10: [76]

Sous les hypothèses (5.24)-(5.26), le système (2.17)-(2.19) a une solution unique $[T(x,t), Y(x,t)]$ sur $\mathbb{R} \times \mathbb{R}_+$. Cette solution satisfait:

$$\begin{cases} T,\ Y \in L^\infty_{loc}(\mathbb{R}_+, L^\infty(\mathbb{R})) , \\ 0 \leq Y(x,t) \leq \sup_{x \in \mathbb{R}} Y^0(x) , \\ T(x,t) \geq T_{min} . \bullet \end{cases} \tag{5.28}$$

Nous allons considérer deux applications différentes de ce résultat. Le problème (2.17)-(2.20), dans lequel le terme f_ϵ correspond au vrai terme exponentiel d'Arrhenius, sera appelé le problème $(\mathcal{P}_\epsilon)$. Par ailleurs, nous considèrerons aussi le problème similaire dans lequel nous remplacerons f_ϵ par une fonction f_ϵ^0 ayant une température d'ignition $T_{ig}^0 \in]T_u^0, T_b^0[$ (attention, la normalisation utilisée ici est différente de celle de (2.28)-(2.30)); nous appellerons ce problème modifié le problème $(\mathcal{P}_\epsilon^0)$. Nous supposerons donc que:

$$\begin{cases} f_\epsilon^0 \in \mathcal{C}^3(\mathbb{R}_+, \mathbb{R}_+) , \\ f_\epsilon^0 \text{ est Lipschitzienne sur } \mathbb{R}_+ , \\ \exists T_{ig}^0 \in]T_u^0, T_b^0[, \quad f_\epsilon^0 \equiv 0 \text{ sur } [T_u^0, T_{ig}^0] , \quad f_\epsilon^0 > 0 \text{ sur }]T_{ig}^0, +\infty[, \end{cases} \tag{5.29}$$

et nous noterons $g_\epsilon^0(T) = (T_b^0 - T) f_\epsilon^0(T)$. De plus, pour ce problème $(\mathcal{P}_\epsilon^0)$, nous supposerons que le nombre de Lewis est égal à 1, et que les données initiales (2.18) satisfont:

$$T^0 + Y^0 \equiv T_b^0 \quad \text{sur } \mathbb{R} . \tag{5.30}$$

Le problème $(\mathcal{P}_\epsilon^0)$ se réduit alors à (5.31)-(5.32) ci-dessous:

$$\begin{cases} T_t = g_\epsilon^0(T) + T_{xx} \ , \\ T(x,0) = T^0(x) \ , \\ T(-\infty,t) = T_u^0 = T^0(-\infty) \ , \quad T(+\infty,t) = T_b^0 = T^0(+\infty) \ , \end{cases} \tag{5.31}$$

$$T(.,t) + Y(.,t) \equiv T_b^0 \ . \tag{5.32}$$

Enfin, nous appellerons $(c_\epsilon^0, \mathcal{T}_\epsilon^0)$ la solution d'onde progressive de (5.31), qui vérifie:

$$\begin{cases} c_\epsilon^0(\mathcal{T}_\epsilon^0)_x = g_\epsilon^0(\mathcal{T}_\epsilon^0) + (\mathcal{T}_\epsilon^0)_{xx} \ , \\ \mathcal{T}_\epsilon^0(-\infty) = T_u^0 \ , \quad \mathcal{T}_\epsilon^0(+\infty) = T_b^0 \ , \end{cases} \tag{5.33}$$

et dont l'existence et l'unicité découle du Théorème 1 ((5.33) est l'équivalent de (2.30), écrit ici avec une normalisation différente).

Pour apporter une réponse au paradoxe de la difficulté de la frontière froide, comme nous l'avons dit au début de ce paragraphe, nous devons montrer que, sous les hypothèses adéquates, la solution du problème $(\mathcal{P}_\epsilon)$ reste proche de la solution stationnaire $\mathcal{T}_\epsilon^0$. C'est l'objet des deux résultats suivants, dont le premier décrit le comportement asymptotique pour les grandes valeurs de t de la solution de $(\mathcal{P}_\epsilon^0)$, et montre que la solution d'onde progressive $\mathcal{T}_\epsilon^0$ est stable:

Théorème 11: [78]

On suppose que les hypothèses (5.24)-(5.25) et (5.29)-(5.30) sont vérifiées, et que $\mathcal{L} = 1$. Soit $(T_\epsilon^0, Y_\epsilon^0)$ l'unique solution du problème $(\mathcal{P}_\epsilon^0)$ (i.e. (5.31)-(5.32)), et soit $(c_\epsilon^0, \mathcal{T}_\epsilon^0)$ l'unique solution de (5.33). On suppose de plus que $T_u^0 \le T^0(x) \le T_b^0$ pour tout $x \in I\!R$, et que la limite $\displaystyle\lim_{x \to -\infty} \frac{T^0(x)}{\exp(c_\epsilon^0 x)}$ existe dans $I\!R_+^*$. Alors, il existe un réel x_0 et deux constantes positives K et r telles que:

$$|T_\epsilon^0(x,t) - \mathcal{T}_\epsilon^0(x + x_0 + ct)| \le K e^{-rt} \quad \forall x, t \ . \bullet \tag{5.34}$$

Pour utiliser ce résultat, nous allons supposer que la fonction f_ϵ^0 que nous avons introduite pour le problème $(\mathcal{P}_\epsilon^0)$ est telle que, dans la limite des grandes énergies d'activation, les fonctions f_ϵ et f_ϵ^0 "sont proches". On suppose donc que les familles de fonctions (f_ϵ) et (f_ϵ^0) sont telles que *(i)* pour tout $\epsilon > 0$, f_ϵ satisfait (5.26), *(ii)* pour tout $\epsilon > 0$ f_ϵ^0 satisfait (5.29), et que *(iii)* il existe une suite (δ_ϵ) vérifiant $\lim_{\epsilon \to 0} \delta_\epsilon = 0$ telle que, pour tout $T > 0$:

$$|f_\epsilon^0(T) - f_\epsilon(T)| \le \delta_\epsilon \ . \tag{5.35}$$

Le résultat suivant montre alors que les solutions des problèmes $(\mathcal{P}_\epsilon)$ et $(\mathcal{P}_\epsilon^0)$ restent proches:

Proposition 12: [78]

On suppose que les hypothèses des Théorèmes 10 et 11 sont vérifiées, ainsi que (5.26). Soit (T_ϵ, Y_ϵ) la solution de $(\mathcal{P}_\epsilon)$, et $(T_\epsilon^0, Y_\epsilon^0)$ la solution de $(\mathcal{P}_\epsilon^0)$. Il existe deux suites (δ'_ϵ) et (t_ϵ) avec $\lim\limits_{\epsilon \to 0} \delta'_\epsilon = 0$ et $\lim\limits_{\epsilon \to 0} t_\epsilon = +\infty$, telles que:

$$\|T_\epsilon(.,t) - T_\epsilon^0(.,t)\|_\infty + \|Y_\epsilon(.,t) - Y_\epsilon^0(.,t)\|_\infty \leq \delta'_\epsilon, \quad \forall t \leq t_\epsilon \; . \bullet \tag{5.36}$$

L'ensemble de ces résultats fournit donc la réponse attendue au paradoxe de la difficulté de la frontière froide. Sous les hypothèses précédentes, et si ϵ est assez petit, la solution du problème $(\mathcal{P}_\epsilon)$ (avec le terme d'Arrhenius) reste proche pendant un temps très long de la solution du problème $(\mathcal{P}_\epsilon^0)$ (avec température d'ignition); mais cette dernière tend rapidement (exponentiellement) vers la solution d'onde progressive $\mathcal{T}_\epsilon^0$. On a donc montré que *la solution du modèle avec terme d'Arrhenius se comporte comme une onde progressive pendant une période qui est d'autant plus longue que l'énergie d'activation est grande (i.e. que ϵ est petit).*

5.2. Pour les numériciens

Les apports des études mathématiques aux études numériques en théorie de la combustion sont également nombreux: ils n'incluent pas seulement des résultats d'analyse numérique proprement dite, puisqu'il arrive aussi qu'un résultat mathématique suggère un algorithme numérique. Nous donnons ci-dessous quelques exemples de ces apports.

5.2.1. Analyse numérique du modèle isobare instationnaire plan

Le premier exemple concerne un résultat de convergence pour l'approximation numérique du problème isobare instationnaire plan. Pour simplifier l'exposé, nous considèrerons ici au lieu de (3.17)-(3.20) le problème plus simple (5.31).

Considérons donc la résolution numérique du problème (5.31), en utilisant un maillage uniforme en espace et en temps, sur un intervalle borné $[-a, a]$, avec, pour éviter des restrictions trop importantes sur le pas de temps tout en décrivant de façon précise l'évolution de la solution, un schéma d'intégration numérique implicite pour le terme de diffusion et explicite pour le terme de réaction (nous renvoyons le lecteur intéressé à [9], [58] pour une discussion détaillée sur le choix des schémas d'intégration et des pas de temps, en fonction des divers temps caractéristiques des phénomènes physiques). De plus, nous utiliserons des conditions aux limites de Neumann aux extrémités $-a$ et $+a$ du domaine de calcul. Le schéma utilisé, s'écrit alors:

$$\frac{T_j^{n+1} - T_j^n}{\Delta t} = \frac{T_{j-1}^{n+1} - 2T_j^{n+1} + T_{j+1}^{n+1}}{\Delta x^2} + g_\epsilon^0(T_j^n) \quad \text{pour } 1 \leq j \leq N \ , \tag{5.37}$$

$$\frac{T_0^{n+1} - T_0^n}{\Delta t} = \frac{T_1^{n+1} - T_0^{n+1}}{\Delta x^2} + g_\epsilon^0(T_0^n) \quad \text{pour } j = 0 \ , \tag{5.38}$$

$$\frac{T_{N+1}^{n+1} - T_{N+1}^n}{\Delta t} = \frac{T_N^{n+1} - T_{N+1}^{n+1}}{\Delta x^2} + g_\epsilon^0(T_{N+1}^n) \quad \text{pour } j = N+1 \ , \tag{5.39}$$

où Δt est le pas de temps, N est le nombre de noeuds intérieurs, et $\Delta x = \dfrac{2a}{N+1}$ est le pas en espace. Pour l'initialisation, on pose bien sûr $T_j^0 = T^0(x_j)$, en notant $x_j = j\Delta x - a$ pour $0 \leq j \leq N+1$.

L'analyse du schéma (5.37)-(5.39) conduit alors au résultat suivant:

Théorème 13: [76]

On suppose que les hypothèses (5.24)-(5.25) et (5.29) sont vérifiées, et qu'il existe deux constantes positives K et δ telles que la donnée initiale T^0 vérifie:

$$\begin{cases} |T^0(x) - T_u^0| + |T^{0\prime}(x)| \leq K \exp(\delta x) \ \ \forall x < 0 \ , \\ |T^0(x) - T_b^0| + |T^{0\prime}(x)| \leq K \exp(-\delta x) \ \ \forall x > 0 \ . \end{cases} \tag{5.40}$$

On note T l'unique solution de (5.31), et (T_j^n, Y_j^n) la solution numérique donnée par le schéma (5.37)-(5.39), et on suppose que le pas de temps Δt satisfait:

$$\Delta t \leq \|f_\epsilon^0\|^{-1}_{L^\infty(T_u^0, T_b^0)} \ . \tag{5.41}$$

Alors, pour tout $t_0 > 0$, il existe une constante positive $M = M(t_0)$ telle que, pout tout j et n vérifiant $0 \leq j \leq N+1$, $0 \leq n\Delta t \leq t_0$, on a:

$$|T(x_j, n\Delta t) - T_j^n| \leq M\left(\Delta t + \Delta x + \frac{\exp(-\delta a)}{\Delta x}\right) \ . \bullet \tag{5.42}$$

Ce résultat montre donc que le schéma (5.37)-(5.39) converge: pour tout j et n vérifiant $0 \leq j \leq N+1$ et $0 \leq n\Delta t \leq t_0$, l'erreur $|T(x_j, n\Delta t) - T_j^n|$ peut être rendue inférieure à tout nombre $\hat{\epsilon}$ choisi à l'avance en prenant Δx et Δt assez petit *et a assez grand.*

Remarque 9: Le terme exponentiel dans (5.42) vient de ce que, sous l'hypothèse (5.40), les quantités $T'(x,t)$ et $T(x,t) - T_u^0$ (resp. $T(x,t) - T_b^0$) décroissent exponentiellement lorsque x tend vers $-\infty$ (resp. $+\infty$), uniformément pour $t \leq t_0$. •

Remarque 10: Le Théorème 13 n'est qu'un exemple parmi plusieurs autres résultats de convergence analogues démontrés à la suite des travaux de Benkhaldoun et Larrouturou [8]: des résultats de convergence similaires existent aussi pour une approximation utilisant un maillage adaptatif, ou pour un schéma d'intégration explicite, ou encore pour le modèle thermo-diffusif multi-dimensionnel (voir [8], [58]). Les preuves de tous ces résultats utilisent de façon cruciale la préservation par l'approximation numérique du "principe du maximum discret" (c'est-à-dire le fait que, sous certaines conditions, le schéma numérique préserve les inégalités $0 \leq Y \leq 1$, $T_{min} \leq T$). •

Remarque 11: Les Théorèmes 11 et 13 montrent que la solution numérique T_j^n converge vers la solution T, qui tend exponentiellement vers $\mathcal{T}(x + ct)$ (à une translation fixe de l'origine près). La méthode de maillage mobile proposée dans [54], [55], [57] permet de construire une solution numérique qui converge non pas vers la solution $T(x,t)$ mais vers une solution translatée $\bar{T} = T(x + X(t), t)$ qui, elle, tend exponentiellement vers la solution stationnaire $\mathcal{T}$ (voir [58]). Ce résultat implique que $X'(t)$ tend exponentiellement vers c: cette méthode de maillage mobile fournit donc aussi la vitesse de la flamme. •

5.2.2. Approximations tronquées de solutions d'onde progressive

Pour le second exemple, nous considérons de nouveau le problème le plus simple de la flamme prémélangée plane stationnaire pour un nombre de Lewis égal à 1. Nous considérerons la normalisation (2.30), et noterons ci-dessous (c, Θ) l'unique couple de $\mathcal{C}^2(\mathbb{R}) \times \mathbb{R}$ qui vérifie (2.30) et la condition de normalisation $\Theta(0) = \theta$.

A part les méthodes utilisant une approche instationnaire, que nous avons considéré au paragraphe précédent, les méthodes utilisées pour le calcul numérique d'une flamme plane stationnaire consistent à résoudre un problème analogue à (2.30) mais posé sur un intervalle borné $[-a, a]$: on cherche par exemple c_a et Θ_a

vérifiant:

$$\begin{cases} c_a(\Theta_a)_x = g_\epsilon^0(\Theta_a) + (\Theta_a)_{xx} \text{ dans } [-a,a] , \\ \Theta_a(-a) = 0 , \quad \Theta_a(0) = \theta , \quad \Theta_a(a) = 1 . \end{cases} \tag{5.43}$$

Les deux résultats suivants montrent l'intérêt de cette approche pour l'approximation de (2.30): la solution tronquée (c_a, Θ_a) converge exponentiellement vers la solution stationnaire (c, Θ):

Théorème 14: [11]

Sous l'hypothèse (5.1), le problème (5.43) a une solution unique (c_a, Θ_a), pour tout $a > 0$. Lorsque a tend vers $+\infty$, c_a converge vers c, et Θ_a converge localement vers Θ. •

Théorème 15: [36]

On suppose que g_ϵ^0 vérifie les hypothèses (5.1), est strictement décroissante au voisinage de 1, et est dérivable à gauche en 1 avec $g_\epsilon^{0\prime}(1) < 0$. Alors, il existe des constantes positives a_0, K et k telles que, pour tout $a \geq a_0$, on a:

$$|c_a - c| \leq Ke^{-ka} , \quad \|\Theta_a - \Theta\|_{C^2(-a,a)} \leq Ke^{-ka} . \bullet \tag{5.44}$$

Remarque 12: D'autres problèmes posés sur un domaine borné sont également utilisés dans les calculs numériques: on remplace par exemple dans (5.43) l'une des conditions aux limites en $-a$ ou a par une condition de Neumann. Il faut noter ici que les résultats précédents d'existence, d'unicité et de convergence exponentielle de la solution tronquée ne sont pas indifféremment valables quelles que soient les conditions aux limites imposées à cette solution en $-a$ et a (voir [36]). •

5.2.3. Ecoulements compressibles multi-espèces

Le troisième exemple concerne le système (2.1) qui décrit l'écoulement d'un mélange gazeux compressible réactif. Ce système n'intervient pas seulement en théorie de la combustion, mais aussi pour décrire des écoulements réactifs transsoniques, supersoniques ou hypersoniques (voir par exemple [25], [27], [34]). Son approximation numérique inclut les difficultés liées à l'approximation des équations d'Euler d'un gaz compressible, qui sont contenues dans (2.1), et fait le plus souvent appel à des méthodes décentrées. Dans ce cadre, la préservation du principe du maximum discret pour les fractions massiques (i.e. le respect par la méthode numérique des inégalités $0 \leq Y \leq 1$ pour toutes les espèces présentes) exige une attention particulière: nous allons voir que, pour les méthodes basées sur l'emploi d'un "solveur de Riemann approché", méthodes très fréquemment utilisées maintenant, la préservation du principe du maximum discret pour les fractions massiques est basé sur une propriété mathématique du problème (5.45) ci-dessous.

Dans un premier temps, considérons pour simplifier le système des équations d'Euler multi-espèces dans lequel on ne conserve que les membres de gauche des équations (2.1):

$$\begin{cases} \rho_\tau + (\rho u)_\xi = 0 \ , \\ (\rho u)_\tau + (\rho u^2 + p)_\xi = 0 \ , \\ E_\tau + [u(E+p)]_\xi = 0 \ , \\ (\rho Y)_\tau + (\rho u Y)_\xi = 0 \ , \end{cases} \tag{5.45}$$

complété par la loi d'état donnant la pression, qui s'écrit d'après (2.2)-(2.3) et la loi de Mayer:

$$p = (\gamma - 1)\left(E - \frac{1}{2}\rho u^2 - Q\rho Y - \rho h^0_{\mathcal{P}}\right) \ . \tag{5.46}$$

Introduisons les notations:

$$W = \begin{pmatrix} \rho \\ \rho u \\ E \\ \rho Y \end{pmatrix} = \begin{pmatrix} W^1 \\ W^2 \\ W^3 \\ W^4 \end{pmatrix} \ , \quad F = \begin{pmatrix} \rho u \\ \rho u^2 + p \\ u(E+p) \\ \rho u Y \end{pmatrix} = \begin{pmatrix} F^1 \\ F^2 \\ F^3 \\ F^4 \end{pmatrix} \ . \tag{5.47}$$

Pour l'approximation numérique de (5.45) (qui peut s'écrire $W_\tau + F_\xi = 0$), nous considèrerons des schémas conservatifs explicites à trois points s'écrivant:

$$\frac{W_j^{n+1} - W_j^n}{\Delta \tau} + \frac{\phi^n_{j+1/2} - \phi^n_{j-1/2}}{\Delta \xi} = 0 \ , \tag{5.48}$$

où le flux numérique $\phi^n_{j+1/2}$ est évalué en utilisant une "fonction de flux numérique" Φ qui définit le schéma considéré:

$$\phi^n_{j+1/2} = \Phi(W_j^n, W_{j+1}^n) \ . \tag{5.49}$$

Nous écrirons:

$$\phi^n_{j+1/2} = \begin{pmatrix} \phi^{1,n}_{j+1/2} \\ \phi^{2,n}_{j+1/2} \\ \phi^{3,n}_{j+1/2} \\ \phi^{4,n}_{j+1/2} \end{pmatrix} \ . \tag{5.50}$$

La plupart des schémas existants pour les équations d'Euler classiques (pour un seul gaz) peuvent être étendus aux équations d'Euler multi-espèces. Ces généralisations ont été étudiées par plusieurs auteurs, et nous renvoyons par exemple à Abgrall-Montagné [2] et Larrouturou-Fezoui [61] pour l'extension du schéma de Godunov, à Abgrall [1] et Fernandez-Larrouturou [37] pour celle du schéma de Roe, à [2] pour celle du schéma d'Osher, et à [61] pour l'extension du schéma de Van Leer et pour une revue de ces schémas multi-espèces.

L'étude de ces différents schémas du point de vue de la préservation du principe du maximum discret pour les fractions massiques montre que les schémas multi-espèces de Godunov et de Van Leer préservent bien le principe du maximum pour les fraction massiques, mais que ce n'est pas le cas des schémas multi-espèces de Roe et d'Osher (voir [26]). Pourquoi ces deux derniers schémas, reposant sur des solveurs de Riemann approchés, ne respectent-ils pas le principe du maximum que préserve bien le schéma de Godunov, basé sur un solveur de Riemann exact? La réponse à cette question, apportée dans [59], [60], repose sur l'étude mathématique du problème de Riemann pour le système (5.45).

Pour présenter cette réponse, commençons par rappeler les propriétés de base du problème (5.45): il est facile de montrer que les composantes du vecteur des flux F s'expriment en fonction des composantes de W, et que le système $W_\tau + F(W)_\xi = 0$ est hyperbolique: la matrice Jacobienne $\left(\frac{\partial F_i}{\partial W_j}\right)$ a toujours quatre valeurs propres réelles $u-c$, u, u, et $u+c$, où $c = \sqrt{\frac{\gamma p}{\rho}}$ est la vitesse du son, et quatre vecteurs propres réels. De plus, quels que soient les états W_G et W_D, on sait calculer la solution exacte $W^{\mathcal{R}}(\xi,\tau)$ du problème de Riemann:

$$\left\{\begin{array}{l} W_\tau + F_\xi = 0 \quad \text{pour} \quad \xi \in I\!R\,, \quad \tau \geq 0\,, \\ W(\xi,\tau=0) = \left\{\begin{array}{ll} W_G & \text{si } \xi < 0\,, \\ W_D & \text{si } \xi > 0\,. \end{array}\right. \end{array}\right. \tag{5.51}$$

Comme dans le cas des équations d'Euler mono-espèces, cette solution exacte est constituée de quatre états constants, séparés par une 1-onde (choc ou raréfaction), une discontinuité de contact, et une 4-onde (choc ou raréfaction) (voir Figure 7). Un point essentiel est que la fraction massique Y ne varie qu'à travers la discontinuité de contact. Nous renvoyons à [61] pour les démonstrations de ces propriétés.

Figure 7: Solution exacte du problème de Riemann multi-espèces.

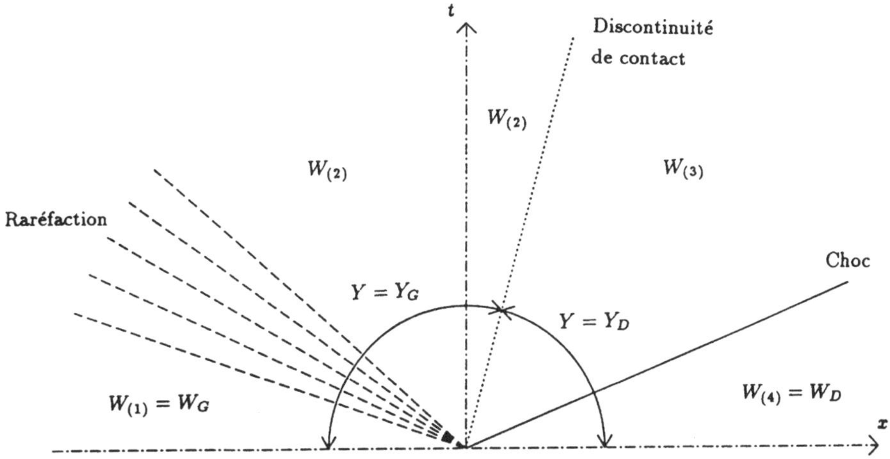

Pour $\sigma \in I\!R$ et $\tau > 0$, nous noterons $\mathcal{W}(\sigma; W_G, W_D)$ la valeur de $W^{\mathcal{R}}(\sigma\tau, \tau)$, qui est indépendante de τ. La propriété essentielle démontrée dans [60] est alors la suivante:

Proposition 16: [60]

Quels que soient les états W_G et W_D, l'égalité suivante a lieu:

$$F^4[\mathcal{W}(0; W_G, W_D)] = F^1[\mathcal{W}(0; W_G, W_D)] \times \begin{cases} Y_G & \text{si } F^1[\mathcal{W}(0; W_G, W_D)] > 0 \,, \\ Y_D & \text{si } F^1[\mathcal{W}(0; W_G, W_D)] < 0 \,. \end{cases} \bullet \tag{5.52}$$

Il apparaît dans [60] que (5.52) est précisément la propriété essentielle que permet au schéma de Godunov (dans lequel $\phi^n_{j+1/2} = F[\mathcal{W}(0; W^n_j, W^n_{j+1})]$) de vérifier le principe du maximum pour les fractions massiques, et que si l'on modifie la quatrième composante $\phi^{4,n}_{j+1/2}$ du flux numérique des schémas de Roe et d'Osher en imposant la relation:

$$\phi^{4,n}_{j+1/2} = \phi^{1,n}_{j+1/2} \times \begin{cases} Y^n_j & \text{si } \phi^{1,n}_{j+1/2} > 0 \,, \\ Y^n_{j+1} & \text{si } \phi^{1,n}_{j+1/2} < 0 \,, \end{cases} \tag{5.53}$$

calquée sur (5.52), alors les schémas multi-espèces de Roe et d'Osher ainsi modifiés préservent le principe du maximum pour les fractions massiques.

La propriété (5.52) permet donc de construire des schémas respectant la positivité de toutes les fractions massiques pour l'approximation numérique de l'écoulement d'un mélange gazeux compressible. On montre de plus dans [60] que ces schémas s'étendent sans difficulté et en conservant ces propriétés pour obtenir un schéma implicite, ou un schéma précis au second ordre, ou pour traiter des écoulements réactifs multi-dimensionnels.

6. CONCLUSION

Cet article montre s'il en est besoin combien la combustion est un domaine pluridisciplinaire, dans lequel les exemples de résultats obtenus par l'une des approches physique, numérique et mathématique qui sont instructifs et féconds pour les autres approches sont nombreux et variés.

Il aurait pleinement atteint son but s'il contribuait à développer les interactions nécessaires entre les communautés des physiciens, des numériciens et des mathématiciens s'intéressant à la combustion, et plus encore s'il contribuait à donner à de jeunes chercheurs le goût et la vocation de devenir spécialistes non seulement de l'une de ces trois approches de la modélisation de la combustion, mais même de l'ensemble si complémentaire de ces trois approches.

BIBLIOGRAPHIE

[1] R. ABGRALL, "Généralisation du schéma de Roe pour le calcul d'écoulements de mélanges de gaz à concentrations variables", La Recherche Aérospatiale, pp. 31-43, **6**, (1988).

[2] R. ABGRALL & J. L. MONTAGNE, "Généralisation du schéma d'Osher pour le calcul d'écoulements de mélanges de gaz à concentrations variables et de gaz réels", La Recherche Aérospatiale, **4**, pp. 1-13, (1989).

[3] M. T. AIMAR, "Etude numérique d'une équation d'évolution non linéaire décrivant l'instabilité thermodiffusive d'un front de flamme", Thèse, Université de Provence, (1982).

[4] D. G. ARONSON & H. F. WEINBERGER, "Nonlinear diffusion in population genetics, combustion and nerve propagation", Partial differential equations and related topics, Lect. Notes in Math., **446**, pp. 5-49, Springer Verlag, New York, (1975).

[5] G. I. BARENBLATT, Ya. B. ZELDOVICH & A. G. ISTRATOV, Prikl. Mekh. Tekh. Fiz., **2**, p. 21, (1962).

[6] A. BAYLISS & B. J. MATKOWSKY, "Fronts, relaxation oscillations, and period doubling in solid fuel combustion", J. Comp. Phys., **71**, pp. 147-168, (1987).

[7] F. BENKHALDOUN, B. DENET & B. LARROUTUROU, "Numerical investigation of the extinction limit of curved flames", Comb. Sci. Tech., **64**, (4-6), pp. 187-198, (1989).

[8] F. BENKHALDOUN & B. LARROUTUROU, "Numerical analysis of the two-dimensional thermo-diffusive model for flame propagation", Mod. Math. et Anal. Num., **22**, (4), pp. 535-560, (1988).

[9] F. BENKHALDOUN & B. LARROUTUROU, "A finite-element adaptive investigation of curved stable and unstable flame front", Comp. Meth. Appl. Mech. Eng., **76**, pp. 119-134, (1989).

[10] H. BERESTYCKI, L. CAFFARELLI & L. NIRENBERG, à paraître.

[11] H. BERESTYCKI & B. LARROUTUROU, "A semilinear elliptic equation in a strip arising in a two-dimensional flame propagation model", J. für Reine und Angewandte Mathematik, **396**, pp. 14-40, (1989).

[12] H. BERESTYCKI & B. LARROUTUROU, "Quelques aspects mathématiques de la propagation des flammes prémélangées", Nonlinear partial differential equations and their applications, Collège de France seminar, Brezis & Lions eds., Research Notes in Mathematics, Pitman-Longman, London, à paraître, (1990).

[13] H. BERESTYCKI et B. LARROUTUROU, "On planar traveling-front solutions of reaction-diffusion problems", Rapport INRIA, à paraître, (1990).

[14] H. BERESTYCKI & B. LARROUTUROU, "Mathematical modelling of planar flame propagation", Research Notes in Mathematics, Pitman-Longman, London, à paraître, (1990).

[15] H. BERESTYCKI, B. LARROUTUROU & P. L. LIONS, "Multi-dimensional traveling wave solutions of an equation describing premixed flame propagation", Arch. Rat. Mech. Anal., **111**, pp. 33-49, (1990).

[16] H. BERESTYCKI, B. LARROUTUROU & P. L. LIONS, "Stationary solutions of the thermo-diffusive model of flame propagation for non unit Lewis number", à paraître.

[17] H. BERESTYCKI, B. LARROUTUROU & L. NIRENBERG, "A nonlinear elliptic problem describing the propagation of a curved premixed flame", Mathematical modelling in combustion and related topics, Brauner & Schmidt-Lainé eds., pp. 11-28, NATO ASI Series E, Nijhoff, Doordrecht, (1988).

[18] H. BERESTYCKI, B. LARROUTUROU & J. M. ROQUEJOFFRE, "Mathematical investigation of the cold boundary difficulty in flame propagation theory", Dynamical issues in combustion, IMA Volume, Springer Verlag, à paraître, (1990).

[19] H. BERESTYCKI, B. LARROUTUROU & J. M. ROQUEJOFFRE, "Stability analysis of curved thermo-diffusive flames - Part I: Linear analysis", à paraître.

[20] H. BERESTYCKI, B. NICOLAENKO & B. SCHEURER, "Traveling wave solutions to combustion models and their singular limits", SIAM J. Math. Anal., **16**, (6), pp. 1207-1242, (1985).

[21] H. BERESTYCKI & L. NIRENBERG, "Monotonicity, symmetry and antisymmetry of solutions of semilinear elliptic equations", J. Geom. and Phys., (special issue dedicated to I. M. Gelfand), à paraître.

[22] H. BERESTYCKI & L. NIRENBERG, "Some qualitative properties of solutions of semilinear elliptic equations in cylindrical domains", Analysis et Cetera, Rabinowitz Zehnder eds., pp. 115-164, Academic Press, New-York, (1990).

[23] H. BERESTYCKI & G. I. SIVASHINSKY, "Flame extinction by periodic flow field", à paraître.

[24] J. D. BUCKMASTER & G. S. S. LUDFORD, "Theory of laminar flames", Cambridge Univ. Press, Cambridge, (1982).

[25] G. V. CANDLER & R. W. MacCORMACK, "The computation of hypersonic flows in chemical and thermal nonequilibrium", Paper N° 107, Third National Aero-Space Plane Technology Symposium, (1987).

[26] D. CHARGY, R. ABGRALL, L. FEZOUI & B. LARROUTUROU, "Comparisons of several numerical schemes for multi-component one-dimensional flows", Rapport INRIA, à paraître.

[27] D. CHARGY, A. DERVIEUX & B. LARROUTUROU, "Upwind adaptive finite-element investigations of two-dimensional transonic reactive flows", Int. J. Num. Meth. Fluids, à paraître.

[28] P. CLAVIN, "Dynamic behavior of premixed flame fronts in laminar and turbulent flows", Prog. Energ. Comb. Sci., **11**, p. 1, (1985).

[29] P. CLAVIN & A. LIÑAN, "Theory of gaseous combustion", Non equilibrium cooperative phenomena in physics and related fields, Velarde éd., NATO ASI Series, Physics B, **116**, Plenum, (1983).

[30] P. CLAVIN, P. PELCE & L. HE, "One-dimensional vibratory instability of planar flames propagating in tubes", à paraître.

[31] G. DARRIEUS, "Propagation d'un front de flamme", non publié, (1938).

[32] B. DENET, "Simulations numériques d'instabilités de fronts de flammes", Thesis, Université de Provence, Marseille, (1988).

[33] A. DERVIEUX, B. LARROUTUROU & R. PEYRET, "On some adaptive numerical approaches of thin flame propagation problems", Comp. Fluids, **17**, (1), pp. 39-60, (1989).

[34] J. A. DESIDERI, N. GLINSKY & E. HETTENA, "Hypersonic reactive flow computations", Computers and Fluids, à paraître.

[35] H. A. DWYER, R. J. KEE & B. R. SANDERS, "Calculation of complex combustion waves with adaptive grids", Seventh Int. Colloquium on Gas Dynanics and Reactive Systems, AIAA, (1978).

[36] R. FERRETTI & B. LARROUTUROU, "Error estimates for truncated domain approximations of travelling-front solutions of a combustion model", à paraître.

[37] G. FERNANDEZ & B. LARROUTUROU, "Hyperbolic schemes for multi-component Euler equations", Nonlinear hyperbolic equations - Theory, numerical methods and applications, Ballmann & Jeltsch eds., pp. 128-138, Notes on Numerical Fluid Mechanics, **24**, Vieweg, Braunschweig, (1989).

[38] M. GHILANI & B. LARROUTUROU, "Upwind computation of steady planar flames with complex chemistry", Mod. Math. et Anal. Num., à paraître.

[39] A. F. GHONIEM, "Dynamics of flame propagation in a turbulent flow field", Lecture in Appl. Math., **22**, pp. 161-175, (1985).

[40] V. GIOVANGIGLI, "Mathematical analysis of non adiabatic planar laminar flames", SIAM J. Math. Anal., à paraître.

[41] L. GLANGETAS, "Etude d'une limite singulière d'un modèle intervenant en combustion", à paraître.

[42] H. GUILLARD, B. LARROUTUROU & N. MAMAN, "Etude numérique des instabilités cellulaires d'un front de flamme par une méthode pseudo-spectrale", Rapport INRIA 721, (1987).

[43] W. E. JOHNSON, "On a first-order boundary value problem for laminar flame theory", Arch. Rat. Mech. Anal., **13**, pp. 46-54, (1963).

[44] W. E. JOHNSON & W. NACHBAR, "Laminar flame theory and the steady linear burning of a monopropellant", Arch. Rat. Mech. Anal., **12**, pp. 58-91, (1963).

[45] G. JOULIN, "Existence, stabilité et structuration des flammes prémélangés", Thesis, Université de Poitiers, (1979).

[46] G. JOULIN & P. CLAVIN, "Analyse asymptotique des conditions d'extinction des flammes laminaires", Acta Astron., **3**, pp. 223-240, (1976).

[47] Ja. I. KANEL', "Stabilization of solutions of the Cauchy problem for equations encountered in combustion theory", Mat. Sbornik, **59**, pp. 245-288, (1962).

[48] Ja. I. KANEL', "On steady state solutions to systems of equations arising in combustion theory", Dokl. Akad. Nauk USSR, **149**, (2), pp. 367-369, (1963).

[49] R. J. KEE & J. A. MILLER, "Computational modelling of flame structure", Sandia National Laboratory Report SAND 83–8235, (1983).

[50] R. J. KEE, L. R. PETZOLD, M. D. SMOOKE & J. F. GRCAR, "Implicit methods in combustion and chemical kinetics modeling", Multiple time scales, Brackbill & Cohen eds., pp. 113-144, Academic Press, New-York, (1985).

[51] R. J. KEE, J. WARNATZ & J. A. MILLER, "A Fortran computer code package for the evaluation of gas–phase viscosities, conductivities, and diffusion coefficients", Sandia National Laboratory Report SAND 83–8209, (1983).

[52] A. N. KOLMOGOROV, I. G. PETROVSKII & N. S. PISKUNOV, "A study of the equation of diffusion with increase in the quantity of matter and its application to a biological problem", Bjul. Moskowskogo Gos. Univ., **17**, pp. 1-26, (1937).

[53] L. LANDAU, Acta Physicochim. URSS, **19**, p. 77, (1944).

[54] B. LARROUTUROU, "Utilisation de maillages adaptatifs pour la simulation de flammes monodimensionnelles instationnaires", Numerical simulation of combustion phenomena, Glowinski Larrouturou & Temam eds., pp. 300-312,

Lecture Notes in Physics, **241**, Springer Verlag, Heidelberg, (1985).

[55] B. LARROUTUROU, "Adaptive numerical methods for unsteady flame propagation", Combustion and chemical reactors, Ludford éd., Lect. in Appl. Math., **24**, (2), pp. 415-435, AMS, Providence, (1986).

[56] B. LARROUTUROU, "The equations of one-dimensional unsteady flame propagation: existence and uniqueness", SIAM J. of Math. Anal., **19**, (1), pp. 32-59, (1988).

[57] B. LARROUTUROU, "A conservative adaptive method for unsteady flame propagation", SIAM J. Sci. Stat. Comp., **10**, (4), pp. 742-755, (1989).

[58] B. LARROUTUROU, "Introduction to combustion modelling", Springer series in computational physics, à paraître, (1990).

[59] B. LARROUTUROU, "Problème de Riemann et approximation numérique pour les équations d'Euler multi-espèces", C. R. Acad. Sci. Paris, **309**, Série I, pp. 893-896, (1989).

[60] B. LARROUTUROU, "How to preserve the mass fraction positivity when computing compressible multi-component flows", Rapport INRIA 1080, (1989).

[61] B. LARROUTUROU & L. FEZOUI, "On the equations of multi-component perfect or real gas flow", Nonlinear hyperbolic problems, Carasso Charrier Hanouzet & Joly eds., Lecture notes in mathematics, **1402**, Springer Verlag, Heidelberg, (1989), to appear.

[62] B. LARROUTUROU, A. LIÑAN & J. VEGA, "Asymptotic analysis of steady non adiabatic thermo-diffusive curved flames", non publié.

[63] A. MAJDA, "Equations for low Mach number combustion", Center for Pure and Appl. Math., Rapport 112, Univ. Californie, Berkeley, (1982).

[64] A. MAJDA & J. SETHIAN, "The derivation and numerical solution of the equations for zero Mach number combustion", Comb. Sci. Tech., **42**, pp. 185-205, (1985).

[65] M. MARION, "Etude mathématique d'un modèle de flamme laminaire sans température d'ignition: I - Cas scalaire", Ann. Fac. Sci. Toulouse, **6**, pp. 215-255, (1984).

[66] M. MARION, "Qualitative properties of a nonlinear system for laminar flames without ignition temperature", Nonlinear Analysis, Theory Methods and Applications, **9**, (11), pp. 1269-1292, (1985).

[67] G. H. MARKSTEIN, "Nonsteady flame propagation", Pergamon Press, (1964).

[68] D. M. MICHELSON & G. I. SIVASHINSKY, "Nonlinear analysis of hydrodynamic instability in laminar flames - Part II: Numerical experiments", Acta Astron., **4**, pp. 1207-1221, (1977).

[69] B. N'KONGA, H. GUILLARD & B. LARROUTUROU, "Numerical investigations of the tulip flame instability - Comparisons with experimental results", à paraître.

[70] E. S. ORAN & J. P. BORIS, "Numerical simulation of reactive flows", Elsevier, New-York, (1987).

[71] P. PELCE & P. CLAVIN, "Influence of hydrodynamics and diffusion upon the stability limits of laminar premixed flames", J. Fluid Mech., **124**, pp. 219-237, (1982).

[72] N. PETERS, "Discussion of test problem A", Numerical methods in laminar flame propagation, Peters & Warnatz eds., pp. 1-14, Notes on Numerical Fluid Mechanics, **6**, Vieweg, Braunschweig, (1982).

[73] N. PETERS, "Numerical and asymptotic analysis of systematically reduced reaction schemes for hydrocarbon flames", Numerical simulation of combustion phenomena, Glowinski Larrouturou & Temam eds., pp. 90-109, Lecture Notes in Physics, **241**, Springer Verlag, Heidelberg, (1985).

[74] N. PETERS & J. WARNATZ éds., "Numerical methods in laminar flame propagation", Notes in Num. Fluid. Mech., **6**, Vieweg, Braunschweig, (1982).

[75] N. PETERS & F. A. WILLIAMS, "The asymptotic structure of stoichiometric methane-air flames", Comb. Flame, (1989).

[76] J. M. ROQUEJOFFRE, "Etude mathématique et numérique de la propagation d'une flamme plame instationnaire", Rapport INRIA, (1990), à paraître.

[77] J. M. ROQUEJOFFRE, "Stability analysis of curved thermo-diffusive flames - Part II: Non linear analysis", à paraître.

[78] J. M. ROQUEJOFFRE, L. SAINSAULIEU & D. TERMAN, "On the stability of steady planar premixed flames", à paraître.

[79] M. SERMANGE, "Mathematical and numerical aspects of one-dimensional laminar flame simulation", Appl. Math. Opt., **14**, (2), pp. 131-154, (1986).

[80] K. SESHADRI & N. PETERS, "Asymptotic structure and extinction of

methane-air diffusion flames", Comb. Flame, **73**, pp. 23-44, (1988).

[81] G. I. SIVASHINSKY, "Diffusional-thermal theory of cellular flames", Comb. Sci. Tech., **15**, pp. 137-146, (1977).

[82] G. I. SIVASHINSKY, "Instabilities, pattern formation and turbulence in flames", Ann. Rev. Fluid Mech., **15**, pp. 179-199, (1983).

[83] G. I. SIVASHINSKY, "Instabilities, pattern formation and turbulence in flames", dans "Modélisation des phénomènes de combustion", Direction des Etudes et Recherches d'Electricité de France, Eyrolles, Paris, (1985).

[84] M. D. SMOOKE, J. A. MILLER & R. J. KEE, "Determination of adiabatic flames speeds by boundary value methods", Comb. Sci. Tech., **34**, pp. 79-90, (1983).

[85] M. D. SMOOKE, A. A. TURNBULL, R. E. MITCHELL & D. E. KEYES, "Solution of two-dimensional axisymmetric laminar diffusion flames by adaptive boundary value methods", Mathematical modelling in combustion and related topics, Brauner & Schmidt-Lainé eds., pp. 261-300, NATO ASI Series E, Nijhoff, Doordrecht, (1988).

[86] J. WARNATZ, "Flame propagation and ignition in hydrocarbon-air mixtures up to octane", Numerical simulation of combustion phenomena, Glowinski Larrouturou & Temam eds., pp. 391-404, Lecture notes in physics, **241**, Springer Verlag, Heidelberg, (1985).

[87] J. WARNATZ & W. JÄGER eds., "Complex chemical reaction systems - Mathematical modelling and simulation", Springer series in chemical physics, Springer Verlag, Heidelberg, (1987).

[88] F. A. WILLIAMS, "Combustion theory", second edition, Benjamin Cummings, Menlo Park, (1985).

[89] Ya. B. ZELDOVICH, "On the theory of flame propagation", J. Chem. Phys., **22**, pp. 27-48, (1948).

[90] Ya. B. ZELDOVICH, G. I. BARENBLATT, V. B. LIBROVICH & G. M. MAHVILADZE, "Mathematical theory of combustion and detonation", Nauka, Moscou, (1980).

[91] Ya. B. ZELDOVICH & D. A. FRANK-KAMENETSKII, Acta Phys. Chim., **2**, p. 348, (1938).

INTRODUCTION À L'ÉTUDE DE LA COMBUSTION

N. Peters
Institut für Technische Mechanik
RWTH Aachen — RFA

C. Kennel *
P.S.A. études et recherches
Rue des Fauvelles - BP 16.
92252 La Garenne Colombe — France
* détachée à l'Institut für Technische Mechanik, RWTH Aachen — RFA

Le texte présenté ici est tiré du cours du Pr. Peters traitant de l'ensemble des phénomènes de combustion. Il aboutit de façon progressive à l'étude de la structure des flammes de prémélange, en utilisant les méthodes asymptotiques. Historiquement, celles-ci ont été développées par Frank-Kamenetzkii et Zeldovich dès les années 1930, en faisant l'hypothèse simplificatrice que la combustion est controlée par une seule réaction globale, à énergie d'activation très grande.

Un des buts de cet exposé est de démontrer que l'approche asymptotique peut être généralisée à partir d'une cinétique plus complexe. Pour parvenir à cette fin, il est d'abord nécessaire de réduire au maximum les mécanismes cinétiques complets proposés dans la littérature. L'accent est donc mis sur la présentation des méthodes récentes de réduction des schémas cinétiques et des méthodes asymptotiques.

Le chapitre 1 reprend les notions de base thermodynamiques, permettant d'aborder les phénomènes de combustion. Le chapitre 2 traite de la cinétique, menant jusqu'au schéma réduit à quatre équations pour le méthane. Au cours des chapitres 3 et 4, qui abordent la combustion des mélanges homogènes, sont présentés les premiers raisonnements utilisant les méthodes asymptotiques. Celles-ci sont développées au chapitre 5 pour une flamme de prémélange, à partir d'une seule réaction à grande énergie d'activation. Ceci correspond au cas étudié par Frank-Kamenetzkii et Zeldovich. La structure de flamme alors mise en évidence comprend une zone de réaction très fine dans laquelle le fuel est consommé, précédée d'une zone de préchauffage chimiquement inerte. Ce résultat est généralisé au chapitre 6 à l'aide du schéma réduit pour une flamme de méthane stœchiométrique. Une zone d'oxydation de CO et H_2 d'épaisseur plus importante que la zone de réaction et située en amont de celle-ci est mise en évidence.

Table des matières

1. Thermodynamique de la combustion 124

1.1 Conservation de la masse dans un système chimique réactif 124

Fractions molaires et fractions massiques 124

Relations stœchiométriques 125

La fraction de mélange 126

1.2 Conservation de l'énergie durant la combustion 126

La température adiabatique de flamme 127

Relation de couplage 128

2. Les réactions cinétiques 130

2.1 La vitesse de réaction 130

Différents types de réactions élémentaires 131

La dépendance en température des constantes de vitesse des réactions 132

La dépendence en pression des constantes de vitesse des réactions 132

2.2 Schéma réactionnel complexe 135

L'oxydation de l'hydrogène 135

L'oxydation du monoxyde de carbone 136

L'oxydation du méthane 137

2.3 Réduction de schéma cinétique 137

Stationnarité des composants chimiques 137

Réduction du schéma cinétiques pour le système H_2/O_2 139

Méthode de réduction à partir des chaînes principales 141

Mécanisme réduit pour le méthane 142

3. Explosion dans les systèmes homogènes 145

Equations de base pour la combustion homogène à volume constant 145

Théorie de l'explosion thermique adiabatique 146

Effet des pertes de chaleur 147

4. Le réacteur parfaitement mélangé 150

Equations décrivant l'évolution du mélange en fonction du temps 151

Solution stationnaire 153

Allumage et extinction 155

Limites asymptotiques pour les solutions stationnaires 156

5. Flammes laminaires 161

Vitesse de combustion laminaire 161

Calcul numérique de la vitesse de combustion 163

Théorie thermique de la flamme 165

Dérivée de la théorie thermique de la flamme — Développement asymptotique pour de grandes énergies d'activation 168

6. Structure asymptotique d'une flamme méthane-air stœchiométrique 174

1. Thermodynamique De La Combustion

1.1 Conservation de la masse dans un système chimique réactif

Une combustion signifie libération d'énergie interne sous forme de chaleur par transformation des élements mis en jeux. Sous l'action des réactions chimiques, le réactif et l'oxydant sont transformés en produits, tel que le dioxyde de carbone et l'eau. L'état initial et l'état final d'un système homogène, ou localement homogène, peut-être déterminé à partir du principe thermodynamique de la conservation de la masse et de l'énergie. La relation algébrique qui relie alors la température et les concentrations est plus facile à manipuler que les équations différentielles qui décrivent localement l'état d'avancement de la combustion.

Fractions molaires et fractions massiques

Pour décrire la transformation des éléments mis en jeux au cours d'une combustion, nous devons déterminer la concentration des espèces chimiques. Les grandeurs utilisées étant différentes d'une discipline à l'autre, définissons celles qui nous serviront par la suite: Soit un système homogène contenant n espèces chimiques differentes. n_i est le nombre de moles de l'espèce i. Le nombre de moles total n_s est

$$n_s = \sum_{i=1}^{n} n_i \,, \tag{1.1}$$

et la fraction molaire de l'espèce i

$$X_i = \frac{n_i}{n_s} \,. \tag{1.2}$$

M_i étant la masse molaire de l'espèce i. La masse du constituant i est

$$m_i = M_i n_i \,, \tag{1.3}$$

et la fraction massique

$$Y_i = \frac{m_i}{m} \tag{1.4}$$

avec $m = \sum_i m_i$, la masse totale. On peut définir une masse molaire moyenne M telle que

$$m = M n_s \,. \tag{1.5}$$

La relation qui lie la fraction massique et la fraction molaire est

$$Y_i = \frac{M_i}{M} X_i \,. \tag{1.6}$$

Nous introduisons ici la fraction massique des éléments atomiques Z_j, tels que C, O, H, N. n_e éléments sont présents dans le système

$$Z_j = \sum_{i=1}^{n} \frac{a_{ij} M_j}{M_i} Y_i = \frac{M_j}{M} \sum_{i=1}^{n} a_{ij} X_i \qquad (j = 1, 2, \ldots, n_e) \tag{1.7}$$

a_{ij} est le nombre d'atomes j contenus dans la molécule i, et M_j la masse molaire de l'élément atomique j. Au cours de la combustion, il y a conservation des fractions massiques des éléments atomiques Z_i. Ces relations, ne contenants pas de termes chimiques seront d'une précieuse utilité par la suite.

Relations stœchiométriques

Soit une réaction globale stœchiométrique qui met en jeux un fuel et de l'oxygène

$$\nu'_F \mathrm{F} + \nu'_{O_2} \mathrm{O_2} = \nu''_{CO_2} \mathrm{CO_2} + \nu''_{H_2O} \mathrm{H_2O}\,. \tag{1.8}$$

La transformation de ν_1 moles du premier composant correspond à la transformation de ν_i moles de l'espèce i

$$\frac{\mathrm{d}n_i}{\nu_i} = \frac{\mathrm{d}n_1}{\nu_1}, \quad \frac{\mathrm{d}Y_i}{\nu_i M_i} = \frac{\mathrm{d}Y_1}{\nu_1 M_1} \qquad (i = 2, 3, \ldots, n)$$
$$\nu_i = \nu''_i - \nu'_i\,. \tag{1.9}$$

Si on intègre cette deuxième relation pour O_2 et F entre un état donné et l'état initial (représenté par l'indice u) on obtient la relation de couplage entre Y_{O_2} et Y_F

$$\frac{Y_{O_2} - Y_{O_2,u}}{\nu'_{O_2} M_{O_2}} = \frac{Y_F - Y_{F,u}}{\nu'_F M_F}\,. \tag{1.10}$$

Pour un mélange stœchiométrique, on a

$$\left.\frac{X_{O_2,u}}{X_{F,u}}\right|_{st} = \frac{\nu'_{O_2}}{\nu'_F}\,. \tag{1.11}$$

On définit ν, le rapport stœchiométrique tel que

$$\left.\frac{Y_{O_2,u}}{Y_{F,u}}\right|_{st} = \frac{\nu'_{O_2} M_{O_2}}{\nu'_F M_F} = \nu\,. \tag{1.12}$$

La relation de couplage s'écrit alors plus simplement

$$\nu Y_F - Y_{O_2} = \nu Y_{F,u} - Y_{O_2,u}\,. \tag{1.13}$$

La fraction de mélange

Prenons le cas le plus courant où la combustion a lieu entre un flux gazeux constitué du combustible, et un flux gazeux constitué du comburant, ce dernier pouvant être de l'oxygène pur, ou de l'air. Il faut dans tous les cas tenir compte de la présence de gaz inerte. Nous utiliserons par la suite l'indice 1 pour représenter le flux de combustible, et l'indice 2 pour représenter le flux de comburant. La fraction de mélange est définie comme étant le rapport de la masse du combustible sur la masse totale

$$Z = \frac{m_1}{m_1 + m_2} . \tag{1.14}$$

$Y_{F,1}$ étant la fraction massique du fuel dans le flux gazeux de combustible et $Y_{O_2,2}$ la fraction massique de l'oxygène dans le flux de comburant ($Y_{O_2,2} = 0,232$ si le comburant est de l'air), on a avant la combustion

$$\begin{aligned} Y_{F,u} &= Y_{F,1}\, Z , \\ Y_{O_2,u} &= Y_{O_2,2}(1 - Z) . \end{aligned} \tag{1.15}$$

La relation de couplage (1.13) s'écrit

$$Z = \frac{\nu Y_F - Y_{O_2} + Y_{O_2,2}}{\nu Y_{F,1} + Y_{O_2,2}} . \tag{1.16}$$

A la stœchiométrie, cette relation devient, avec (1.12)

$$Z_{st} = \left[1 + \frac{\nu'_{O_2} M_{O_2} Y_{F,1}}{\nu'_F M_F Y_{O_2,2}} \right]^{-1} , \tag{1.17}$$

et la richesse du mélange $\phi = (Y_{F,u}/Y_{O_2,u})/(Y_{F,u}/Y_{O_2,u})_{st}$

$$\phi = \frac{Z}{1 - Z} \frac{(1 - Z_{st})}{Z_{st}} . \tag{1.18}$$

1.2 Conservation de l'énergie durant la combustion

Pour un mélange de gaz parfait, l'enthalpie spécifique (énergie interne) est la somme des enthalpies (énergies internes) de chaque composant

$$h = \sum_{i=1}^{n} Y_i h_i , \qquad u = \sum_{i=1}^{n} Y_i u_i , \tag{1.19}$$

où h_i et u_i sont respectivement l'enthalpie partielle spécifique et l'énergie interne spécifique de l'espèce i, dont l'expression en fonction de la température est

$$h_i = h_{i,\mathrm{ref}} + \int_{T_{\mathrm{ref}}}^{T} c_{pi}^0 \, \mathrm{d}T \,, \qquad u_i = u_{i,\mathrm{ref}} + \int_{T_{\mathrm{ref}}}^{T} c_{vi}^0 \, \mathrm{d}T \,. \tag{1.20}$$

c_{pi}^0, c_{vi}^0 étant les capacités calorifiques spécifiques de l'espèce i, à pression constante et à volume constant respectivement. L'indice ref indique une température de référence, à laquelle l'enthalpie de formation a été déterminée. L'exposant 0 indique que la pression de référence est $p_0 = 1\,\mathrm{atm}$. Pour les gaz parfaits la capacité calorifique est indépendente de la pression. Pour plus de simplicité nous supposorons que nous sommes en présence de gaz parfait, et l'exposant 0 sera omis dans la suite. La capacité calorifique moyenne d'un mélange de gaz parfait est à pression constante

$$c_p = \sum_{i=1}^{n} Y_i \, c_{pi} \,. \tag{1.21}$$

On peut également parler en terme d'enthalpie molaire, ou de capacité calorifique molaire. On a alors

$$H_i = M_i \, h_i \,, \qquad C_{pi} = M_i \, c_{pi} \,. \tag{1.22}$$

L'enthalpie et la capacité calorifique molaire moyenne du mélange s'écrivent

$$C_p = \sum_{i=1}^{n} X_i \, C_{pi} \,, \qquad H = \sum_{i=1}^{n} X_i \, H_i \,. \tag{1.23}$$

L'enthalpie d'une réaction est l'énergie libérée lorsque cette réaction a lieu, pour une température constante. Elle est définie par

$$\Delta H = \sum_{i=1}^{n} \nu_i \, H_i \,. \tag{1.24}$$

ΔH est négatif pour une réaction exothermique, et positif pour une réaction endothermique. Une réaction globale de combustion peut toujours s'écrire de façon à ce que ν_F' soit égal à 1. $-\Delta H$ représente alors l'énergie libérée lorsque une mole de combustible est consommée.

La température adiabatique de flamme

L'énergie interne d'un système augmente lorsqu'une quantité de chaleur lui est apportée, et diminue lorsque le système fournit un travail. Le premier principe de la thermodynamique s'écrit

$$\mathrm{d}q + \mathrm{d}w_R = \mathrm{d}u + p \, \mathrm{d}v = \mathrm{d}h - v \, \mathrm{d}p \,. \tag{1.25}$$

q est la chaleur spécifique et w_R le travail spécifique des forces de frottement. Si nous supposons le système adiabatique (qui n'échange pas de chaleur avec l'extérieur), et les forces de frottement nulles, on a $\mathrm{d}h = 0$, c'est à dire, pour un mélange gazeux

$$h = \left(\sum_{i=1}^{n} Y_i\, h_i\right)_u = \left(\sum_{i=1}^{n} Y_i\, h_i\right)_b . \tag{1.26}$$

L'indice u est relatif aux conditions des gaz frais (avant la combustion) et l'indice b aux conditions des gaz brûles (après la combustion). En utilisant la relation (1.20), on obtient

$$h = \sum_{i=1}^{n}(Y_{i,u} - Y_{i,b})\, h_{i,\mathrm{ref}} = \int_{T_{\mathrm{ref}}}^{T_b} c_{p,b}\,\mathrm{d}T - \int_{T_{\mathrm{ref}}}^{T_u} c_{p,u}\,\mathrm{d}T \tag{1.27}$$

avec

$$c_{p,b} = \sum_{i=1}^{n} Y_{i,b}\, c_{pi}\,, \qquad c_{p,u} = \sum_{i=1}^{n} Y_{i,u}\, c_{pi}\,. \tag{1.28}$$

Relation de couplage

Dans le cas particulier où la combustion complète peut être décrite par la relation (1.9), on peut intégrer celle–ci pour $i = O_2$, CO_2, H_2O

$$Y_{i,u} - Y_{i,b} = (Y_{F,u} - Y_{F,b})\,\frac{\nu_i\, M_i}{\nu_F\, M_F}\,. \tag{1.29}$$

En multipliant les deux membres par $h_{i,\mathrm{ref}}$ et en faisant la somme sur $i = 1, 2, \ldots, n$, on obtient, à l'aide de (1.22) et (1.24)

$$\sum_{i=1}^{n}(Y_{i,u} - Y_{i,b})\, h_{i,\mathrm{ref}} = (Y_{F,u} - Y_{F,b})\,\frac{(-\Delta H)_{\mathrm{ref}}}{\nu_F'\, M_F}\,, \tag{1.30}$$

(on a remplacé $-\nu_F$ par ν_F'). Une simplification importante peut être apportée si on suppose la capacité calorifique constante. Pour la combustion dans l'air, l'azote est une espèce majoritaire, et un calcul approximatif est obtenu en posant $c_p = c_{p,b} = c_{p,u} = 0.30\,\mathrm{cal/g\,K}$. Pour un mélange pauvre, on a combustion complète du fuel ($Y_{F,b} = 0$); avec $Y_{F,u} = Z Y_{F,1}$ et c_p constante, on détermine la température adiabatique de flamme d'après (1.27)

$$T_b(Z) = T_u(Z) + \frac{(-\Delta H)_{\mathrm{ref}}\, Y_{F,1}}{c_p\, \nu_F'\, M_F}\, Z \qquad (Z \le Z_{\mathrm{st}}) \tag{1.31}$$

$T_u(Z)$ est la température du mélange gazeux. Si T_1 et T_2 sont les températures des flux gazeux de combustible et de comburant respectivement, $T_u(Z)$ est une fonction linéaire de la fraction de mélange

$$T_u(Z) = T_2 + Z\,(T_1 - T_2)\,. \tag{1.32}$$

En général, T_1 et T_2 sont peu différents, et le deuxième terme du membre de droite de la relation (1.31) est prédominant. La température adiabatique est alors une fonction linéaire de Z, et atteint son maximum pour $Z = Z_{st}$.

De la même manière, pour un mélange riche on a $Y_{O_2,b} = 0$ et $Y_{O_2,u} = (1 - Z)Y_{O_2,2}$ et la relation (1.9) s'écrit pour $i =$ F, CO_2 et H_2O

$$Y_{i,u} - Y_{i,b} = (Y_{O_2,u} - Y_{O_2,b})\,\frac{\nu_i\,M_i}{\nu_{O_2}\,M_{O_2}}\,, \tag{1.33}$$

et on obtient la relation de couplage

$$T_b(Z) = T_u(Z) + \frac{(-\Delta H)_{\rm ref}\,Y_{O_2,2}}{c_p\,\nu_O'\,M_{O_2}}\,(1 - Z) \qquad (Z \geq Z_{st}) \tag{1.34}$$

A partir de son maximum, pour $Z = Z_{st}$, la température décroit pour Z croissant jusqu'à la valeur T_1 pour $Z = 1$. La figure 1 représente schématiquement l'évolution de la température et des fractions massiques en fonction de Z.

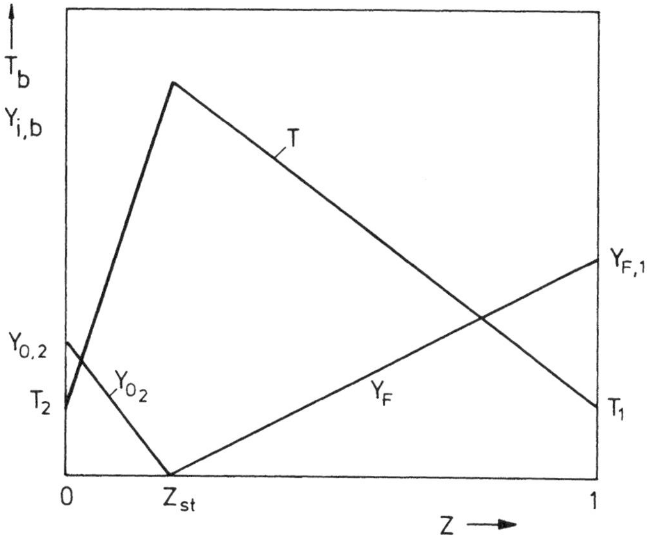

Figure 1:

Evolution de la température et des fractions massiques en fonction de la fraction de mélange.

2. Les Réactions Cinétiques

2.1 La vitesse de réaction

La cinétique chimique a comme principale grandeur la vitesse de réaction. Elle indique le coût molaire d'une réaction par unité de temps. Pour un volume donné, une réaction élémentaire

$$A + B \xrightarrow{k_f} C + D$$

consomme une mole de A et une mole de B par unité de temps, alors qu'une mole de D est produite. La variation au cours du temps du nombre de moles d'une espèce est proportionnelle au nombre de moles de A et au nombre de moles de B contenus dans ce volume

$$\left(\frac{\mathrm{d}C_A}{\mathrm{d}t}\right)_f = -k_f C_A C_B \,. \tag{2.1}$$

où C_i est la concentration molaire de i $C_i = n_i/V = \rho Y_i/M_i$, et k_f un facteur de proportionnalité dépendant de la température (et aussi exceptionnellement de la pression). k_f est la constante de vitesse. Le signe négatif indique que l'espèce A est consommée. De façon plus générale, pour une réaction

$$\nu'_A A + \nu'_B B \longrightarrow \nu''_C C + \nu''_D D$$

on a

$$\frac{1}{\nu_i}\left(\frac{\mathrm{d}C_i}{\mathrm{d}t}\right)_f = \frac{1}{\nu_j}\left(\frac{\mathrm{d}C_j}{\mathrm{d}t}\right)_f = w_f \tag{2.2}$$

avec $\nu_i = \nu''_i - \nu'_i$. w_f est la vitesse de la réaction. Si les espèces C et D sont en présence l'une de l'autre, alors la réaction inverse

$$\nu'_A A + \nu'_B B \longleftarrow \nu''_C C + \nu''_D D$$

peut également avoir lieu. En tenant compte de ce phénomène, la variation globale du nombre de moles de A est donnée par

$$\frac{\mathrm{d}C_A}{\mathrm{d}t} = \left(\frac{\mathrm{d}C_A}{\mathrm{d}t}\right)_f + \left(\frac{\mathrm{d}C_A}{\mathrm{d}t}\right)_b = \nu_A (w_f - w_b) = \nu_A w \tag{2.3}$$

avec

$$w = k_f C_A^{\nu'_A} C_B^{\nu'_B} - k_b C_C^{\nu''_C} C_D^{\nu''_D} \,.$$

(Les indices f et b proviennent des mots anglais forward et backward: vers l'avant et vers l'arrière.)

Différents types de réactions élémentaires

Les réactions bimoléculaires font intervenir deux molécules entre elles, ou une molécule et un atome. Une espèce qui a des électrons de valence libres est appelée espèce radicalaire, ou radical (CH_3, CH_2O, H, etc.)

$$CH_4 + H \longrightarrow CH_3 + H_2$$

$$CH_3 + O \longrightarrow CH_2O + H\,.$$

La plupart des réactions élémentaires sont bimoléculaires. Mais il existe aussi des réactions monomoléculaires et trimoléculaires. (Les réactions faisant intervenir simultanément quatre espèces ou plus sont suffisamment improbables pour pouvoir être négligées.)

Une réaction monomoléculaire est la décomposition spontannée de molécules assez grandes en molécules plus petites. Par exemple, pour le pentane

$$C_5H_{12} \longrightarrow C_4H_9 + CH_3\,.$$

Une réaction trimoléculaire a lieu entre trois espèces de masse molaire assez faible. En général, il n'y a pas production d'une seule molécule, car les énergies liants les trois espèces seraient trop faibles pour qu'une telle molécule ne se redissocie pas aussitôt. En fait, une des trois espèces reste chimiquement inerte, et ne sert qu'à absorber l'énergie libérée lors de la récombinaison des deux autres espèces. Par exemple

$$H + H + H_2 \longrightarrow H_2 + H_2$$

et de façon plus générale

$$A + B + M \xrightarrow{k_f} AB + M$$

$$AB + M \xrightarrow{k_b} A + B + M\,.$$

La vitesse de la réaction est

$$w = k_f\, C_A\, C_B\, C_M - k_b\, C_{AB}\, C_M$$

avec

$$C_M = \sum_{i=1}^{n} z_i\, C_i\,.$$

z_i représente l'efficacité de collision relative à l'espèce i. Pour certaines espèces, l'efficacité z_i est connue

$$z_{H_2} = 1.0\,, \quad z_{CH_4} = 6.5\,, \quad z_{H_2O} = 6.5\,, \quad z_{CO} = 0.75$$
$$z_{O_2} = 0.4\,, \quad z_{CO_2} = 1.5\,, \quad z_{N_2} = 0.4\,, \quad z_{Ar} = 0.35$$

(Référence: J. Warnatz, dans Combustion Chemistry (W.C. Gardiner, Ed.), Springer, 1984.) Lorsque pour une réaction donnée les z_i ne sont pas déterminés, on pose $z_i = 1$, et la concentration C_M devient

$$C_M = \frac{p}{RT}\,.$$

La dépendance en température des constantes de vitesse des réactions

Elle prend la forme suivante

$$\begin{aligned} k_f(T) &= B_f\, T^{n_f} \exp\left(-\frac{E_f}{RT}\right)\,, \\ k_b(T) &= B_b\, T^{n_b} \exp\left(-\frac{E_b}{RT}\right)\,. \end{aligned} \tag{2.4}$$

Lorsqu'une réaction est à l'équilibre, la vitesse de la réaction est égale à la vitesse de la réaction inverse $w_f = w_b$ et on a la relation suivante

$$k_f \prod_{i=1}^{n} C_i^{\nu_i'} = k_b \prod_{i=1}^{n} C_i^{\nu_i''} \tag{2.5}$$

$$\frac{k_f(T)}{k_b(T)} = K_C(T) = \prod_{i=1}^{n} C_i^{\nu_i} \qquad (\nu_i = \nu_i'' - \nu_i') \tag{2.6}$$

$K_C(T)$ est la constante d'équilibre de la réaction. Elle est déterminée à partir des données thermodynamiques des espèces mises en jeux.

La dépendence en pression des constantes de vitesse des réactions

Dans beaucoup de cas, suivant les conditions dans lesquelles on se trouve, une réaction peut être soit bimoléculaire, soit trimoléculaire. C'est le cas par exemple de la réaction

$$CH_3 + H + M \longrightarrow CH_4 + M$$

qui se trouve en concurence avec

$$CH_3 + H \longrightarrow CH_4 .$$

On peut regrouper les deux réactions sous la forme

$$CH_3 + H\ (+M) \longrightarrow CH_4\ (+M) .$$

Lorsque la pression augmente, la vitesse de la réaction de recombination, proportionelle à la densité à la puissance 3, augmente plus rapidement que la vitesse de réaction bimoléculaire qui est proportionelle à la densité au carré. Le modèle de Lindemann prend en considération ce phénomène

$$\begin{aligned} AB + M &\xrightarrow{k_{f1}} (AB)^* + M \\ (AB)^* + M &\xrightarrow{k_{b1}} AB + M \\ (AB)^* &\xrightarrow{k_{f2}} A + B \end{aligned} \tag{2.7}$$

$(AB)^*$ représente une espèce intermédiaire complexe, qui peut réagir suivant deux voies: suivant la réaction inverse ou suivant une réaction monomoléculaire. En supposant que l'espèce $(AB)^*$ est une espèce stationnaire, c'est à dire que $\mathrm{d}C_{(AB)^*}/\mathrm{d}t = 0$, on calcule la concentration de $(AB)^*$

$$C_{(AB)^*} = \frac{k_{f1}\, C_{AB}\, C_M}{k_{f2} + k_{b1}\, C_M} \tag{2.8}$$

et on peut déterminer la vitesse de consommation de AB

$$\frac{\mathrm{d}C_{AB}}{\mathrm{d}t} = -\frac{k_{f2} k_{f1}}{k_{f2} + k_{b1}\, C_M}\, C_{AB}\, C_M . \tag{2.9}$$

La concentration C_M est liée à la pression. La constante de la vitesse de la réaction

$$AB\ (+M) \xrightarrow{k_f} A + B\ (+M)$$

peut s'écrire

$$k_f = -\frac{\mathrm{d}C_{AB}/\mathrm{d}t}{C_{AB}} = \frac{k_{f1}\, k_{f2}\, C_M}{k_{f2} + k_{b1}\, C_M} . \tag{2.10}$$

Les valeurs limites sont

- pour une pression très grande

$$k_f\,(p \to \infty) = k_{f\infty} = \frac{k_{f1}\,k_{f2}}{k_{b1}}\,. \qquad (2.11)$$

$k_{f\infty}$ est indépendent de la pression.

- pour une faible pression

$$k_f\,(p \to 0) = k_{f0}\,C_M = k_{f1}\,C_M\,. \qquad (2.12)$$

La constante de vitesse est alors proportionnelle à la concentration molaire.

On peut exprimer k_f en fonction de k_{f0} et $k_{f\infty}$ en utilisant (2.11) et (2.12)

$$k_f = \frac{k_{f0}\,C_M}{1 + k_{f0}\,C_M / k_{f\infty}} = \frac{k_{f\infty}}{1 + k_{f\infty}/(k_{f0}\,C_M)}\,. \qquad (2.13)$$

La figure 2 représente la constante de vitesse de la réaction

$$CH_3 + H\ (+M) \longrightarrow CH_4\ (+M)$$

en fonction de la pression, pour differentes températures:

pour $p > 1000\,\text{atm}$, on a $k \simeq k_\infty$

pour $p < 50\,\text{atm}$, on a $k = const. \times p$

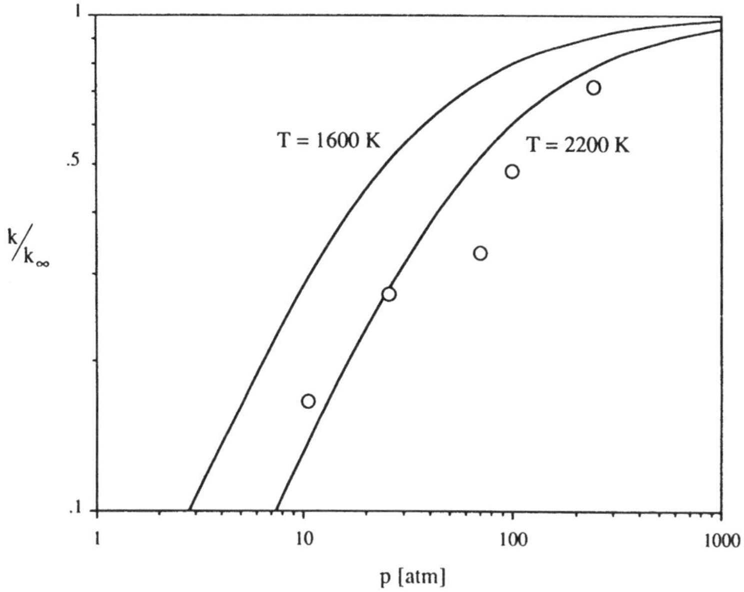

Figure 2:
Evolution de la constante de vitesse de la réaction $CH_3 + H\,(+M) \rightarrow CH_4\,(+M)$ en fonction de la pression.

2.2 Schéma réactionnel complexe

On peut décrire un phénomène de combustion en suivant pas à pas les transformations occasionnées par toutes les réactions élémentaires qui peuvent avoir lieu. Récement, de nombreux schémas cinétiques complets ont ou le jour, permettant cette démarche. Une particularité importante de la combustion est l'enchaînement des réactions élémentaires par l'intermédiaire de radicaux, en particulier H, O, OH.

L'oxydation de l'hydrogène

L'hydrogène est le combustible le plus simple, et par la-même, le mieux connu. La réaction la plus importante du système H_2/O_2 est la réaction de ramification

1 $$H + O_2 \rightleftharpoons OH + O$$

qui à partir d'un radical H, et d'une espèce stable O_2, forment deux radicaux: OH et O. Ceux-ci sont ensuite très rapidement transformés en radical H suivant les réactions 2, 3 et 4, qui constituent une chaîne "¡d'échange"¿ de radicaux

2	$O + H_2 \rightleftharpoons OH + H$
3	$OH + H_2 \rightleftharpoons H_2O + H$
4	$OH + OH \rightleftharpoons H_2O + O$

Il est intéressant de remarquer que les réactions 2, 3 et 4 sont linéairement dépendantes. Si on ajoute la réaction 2 à la réaction 4, on obtient la réaction 3. Lorsque le système est proche de l'état d'équilibre, on peut établir l'équilibre partiel ente H et OH indifféremment à partir de la réaction 3 ou des réactions 2 et 4. On a la relation suivante entre les constantes d'équilibre

$$K_{C3} = K_{C2} K_{C4} .$$

Cela signifie aussi que les constantes de vitesse ne sont pas indépendantes entre elles. La chaîne de rupture, par laquelle les radicaux vont disparaître, a lieu, en présence d'oxygène, d'abord à partir de la réaction 5, suivie de 6, 7 et 8

5	$H + O_2 + M \rightarrow HO_2 + M$
6	$H + HO_2 \rightarrow OH + OH$
7	$H + HO_2 \rightarrow H_2 + O_2$
8	$OH + HO_2 \rightarrow H_2O + O_2$

Dans des conditions de flammes riches, lorques O_2 ne se trouve pas en quantité suffisante pour que la réaction 5 ait lieu, les réactions de recombinaison 24 et 25 deviennent importantes

24	$OH + H + M \rightarrow H_2O + M$
25	$H + H + M \rightarrow H_2 + M$

L'oxydation du monoxyde de carbone

Il n'y a pratiquement qu'une seule réaction importante qui permet d'oxyder le CO en CO_2

9	$CO + OH \rightleftharpoons CO_2 + H$

Cela signifie que le CO pur ne peut être brûlé correctement que s'il se trouve en présence d'hydrogène, qui permet la formation du radical OH. L'air humide suffit pour obtenir un milieu favorable à la combustion du CO. En présence d'air totalement sec, les réactions telles que

$$CO + O + M \rightarrow CO_2 + M$$
$$CO + O_2 \rightarrow CO_2 + O$$

sont trop lentes pour oxyder le CO.

L'oxydation du méthane

Nous nous interesserons ici plus particulièrement à l'oxydation du méthane, qui représente l'espèce hydrocarbonnée la plus simple. Le mécanisme d'oxydation du méthane donné tableau 1 est un mécanisme très réduit, ne comprenant que la chaîne en C_1, c'est à dire la chaîne CH_4, CH_3, CH_2O, CHO, CO. Un tel mécanisme est valable pour les flammes de méthane pauvres ou stœchiométriques. Pour des conditions plus riches, les réactions de recombinaison $CH_3 + CH_3$ qui forment C_2H_5 et C_2H_5 deviennent plus importantes. La chaîne en C_2 ne peut alors plus être négligée. Elle peut ammener à la formation d'hydrocarbures plus lourds, jusqu'à la formation de suies.

2.3 Réduction de schéma cinétique

Il existe déjà de nombreux mécanismes réduits ne comportant que les réactions importantes pour une utilisation courante. Il est cependant possible de réduire encore le nombre de réactions, sans pour autant renoncer aux réactions élémentaires ou aux espèces importantes. Pour cela, deux hypothèses peuvent être utilisées: l'hypothèse de la quasi-stationnarité des espèces, et l'hypothèse des équilibres partiels.

Stationnarité des composants chimiques

La plupart des espèces intermédiaires, en particulier les radicaux, sont consommés très rapidement dès qu'ils apparaissent. Leur concentration reste donc très faible dans la flamme. Pour ces espèces on peut faire l'hypothèse de stationnarité. Cela signifie que les réactions de formation de ces espèces sont beaucoup plus lentes que les réactions de consommation. Soit une espèce i de concentration C_i, on peut écrire, conformément à la relation (2.3)

$$L(C_i) = \sum_{k=1}^{r} \nu_{ik}\, w_k \qquad (i = 1, 2, \ldots, n) \tag{2.14}$$

où k représentent les numéros des réactions du mécanisme considéré. $L(C_i)$ est l'opérateur représentant la variation dans le temps de la concentration C_i

$$L(C_i) = \frac{\mathrm{d}C_i}{\mathrm{d}t}\,.$$

Pour une espèce stationnaire, le terme $\mathrm{d}C_i/\mathrm{d}t$ est négligeable devant le terme de réaction, et on a

$$\sum_{k=1}^{r} \nu_{ik}\, w_k = 0\,.$$

Number	Reaction			B	n	E
$1f$	$H + O_2$	$\rightarrow$	$O + OH$	2.0×10^{14}	0.0	16800
$1b$	$O + OH$	$\rightarrow$	$O_2 + H$	1.575×10^{13}	0.0	690
$2f$	$O + H_2$	$\rightarrow$	$OH + H$	1.8×10^{10}	1.0	8826
$2b$	$OH + H$	$\rightarrow$	$O + H_2$	8.0×10^{9}	1.0	6760
$3f$	$H_2 + OH$	$\rightarrow$	$H_2O + H$	1.17×10^{9}	1.3	3626
$3b$	$H_2O + H$	$\rightarrow$	$H_2 + OH$	5.09×10^{9}	1.3	18588
$4f$	$OH + OH$	$\rightarrow$	$H_2O + O$	6.0×10^{8}	1.3	0
$4b$	$H_2O + O$	$\rightarrow$	$OH + OH$	5.9×10^{9}	1.3	17029
5	$H + O_2 + M^a$	$\rightarrow$	$HO_2 + M^a$	2.3×10^{18}	−0.8	0
6	$H + HO_2$	$\rightarrow$	$OH + OH$	1.5×10^{14}	0.0	1004
7	$H + HO_2$	$\rightarrow$	$H_2 + O_2$	2.5×10^{13}	0.0	700
8	$OH + HO_2$	$\rightarrow$	$H_2O + O_2$	2.0×10^{13}	0.0	1000
$9f$	$CO + OH$	$\rightarrow$	$CO_2 + H$	1.51×10^{7}	1.3	−758
$9b$	$CO_2 + H$	$\rightarrow$	$CO + OH$	1.57×10^{9}	1.3	22337
$10f$	$CH_4\ (+ M)^b$	$\rightarrow$	$CH_3 + H\ (+ M)^b$	6.3×10^{14}	0.0	104000
$10b$	$CH_3 + H\ (+ M)^b$	$\rightarrow$	$CH_4\ (+ M)^b$	5.20×10^{12}	0.0	−1310
$11f$	$CH_4 + H$	$\rightarrow$	$CH_3 + H_2$	2.2×10^{4}	3.0	8750
$11b$	$CH_3 + H_2$	$\rightarrow$	$CH_4 + H$	9.57×10^{2}	3.0	8750
$12f$	$CH_4 + OH$	$\rightarrow$	$CH_3 + H_2O$	1.6×10^{6}	2.1	2460
$12b$	$CH_3 + H_2O$	$\rightarrow$	$CH_4 + OH$	3.02×10^{5}	2.1	17422
13	$CH_3 + O$	$\rightarrow$	$CH_2O + H$	6.8×10^{13}	0.0	0
14	$CH_2O + H$	$\rightarrow$	$HCO + H_2$	2.5×10^{13}	0.0	3991
15	$CH_2O + OH$	$\rightarrow$	$HCO + H_2O$	3.0×10^{13}	0.0	1195
16	$HCO + H$	$\rightarrow$	$CO + H_2$	4.0×10^{13}	0.0	0
17	$HCO + M$	$\rightarrow$	$CO + H + M$	1.6×10^{14}	0.0	14700
18	$CH_3 + O_2$	$\rightarrow$	$CH_3O + O$	7.0×10^{12}	0.0	25652
19	$CH_3O + H$	$\rightarrow$	$CH_2O + H_2$	2.0×10^{13}	0.0	0
20	$CH_3O + M$	$\rightarrow$	$CH_2O + H + M$	2.4×10^{13}	0.0	28812
21	$HO_2 + HO_2$	$\rightarrow$	$H_2O_2 + O_2$	2.0×10^{12}	0.0	0
$22f$	$H_2O_2 + M$	$\rightarrow$	$OH + OH + M$	1.3×10^{17}	0.0	45500
$22b$	$OH + OH + M$	$\rightarrow$	$H_2O_2 + M$	9.86×10^{14}	0.0	−5070
$23f$	$H_2O_2 + OH$	$\rightarrow$	$H_2O + HO_2$	1.0×10^{13}	0.0	1800
$23b$	$H_2O + HO_2$	$\rightarrow$	$H_2O_2 + OH$	2.86×10^{13}	0.0	32790
24	$OH + H + M^a$	$\rightarrow$	$H_2O + M^a$	2.2×10^{22}	−2.0	0
25	$H + H + M^a$	$\rightarrow$	$H_2 + M^a$	1.8×10^{18}	−1.0	0

Tableau 1:

Mécanisme simplifié décrivant l'oxydation du méthane.

Réduction du schéma cinétique pour le système H_2/O_2

Le schéma initial prend en compte les réactions 1 à 8 du tableau 1. Faisons d'abord l'hypothèse que seuls les radicaux OH, O et HO_2 sont stationnaires. Pour O et OH, nous avons déjà vu qu'ils étaient très rapidement consommés par les réactions 2, 3 et 4, après avoir été formés par la réaction 1. De la même manière, la stationnarité de HO_2 est basée sur le fait que la réaction de formation 5 est plus lente que les réactions de consommation 6, 7 et 8. On peut exprimer, pour chacune des espèces, l'opérateur $L(C_i)$ en fonction des termes de réaction. On obtient le système suivant, entre les vitesses de réaction des réactions 1 à 8

$$L(C_H) = -w_1 + w_2 + w_3 - w_5 - w_6 - w_7 \tag{2.15}$$

$$L(C_{OH}) = w_1 + w_2 - w_3 - 2w_4 + 2w_6 - w_8 = 0 \tag{2.16}$$

$$L(C_O) = w_1 - w_2 + w_4 = 0 \tag{2.17}$$

$$L(C_{H_2}) = -w_2 - w_3 + w_7 \tag{2.18}$$

$$L(C_{O_2}) = -w_1 - w_5 + w_7 + w_8 \tag{2.19}$$

$$L(C_{H_2O}) = w_3 + w_4 + w_8 \tag{2.20}$$

$$L(C_{HO_2}) = w_5 - w_6 - w_7 - w_8 = 0 \tag{2.21}$$

Les trois relations de stationnarité nous permettent d'éliminer au moins trois vitesses de réaction, à savoir w_2, w_3 et w_7. Il faut noter ici que le choix des réactions à éliminer est arbitraire. Il est possible d'effectuer le même exercice en éliminant w_6 au lieu de w_7 par exemple. Les réactions de formation des espèces stationnaires ne doivent toutefois pas être éliminées, pour ne pas déséquilibrer le système. On peut écrire w_2, w_3 et w_7 en fonction des autres termes de réaction

(2.17) : $$w_2 = w_1 + w_4$$

(2.16) : $$w_3 = 2w_1 - w_4 + 2w_6 - w_8$$

(2.21) : $$-w_7 = w_6 + w_8 - w_5$$

et les remplacer dans (2.15), (2.18), (2.19) et (2.20). On obtient

$$L(C_H) = 2w_1 - 2w_5 + 2w_6 \tag{2.22}$$

$$L(C_{H_2}) = -3w_1 + w_5 - 3w_6 \tag{2.23}$$

$$L(C_{O_2}) = -w_1 - w_6 \tag{2.24}$$

$$L(C_{H_2O}) = 2w_1 + 2w_6\,. \tag{2.25}$$

On remarque alors que les vitesses de réaction peuvent être regroupées de la façon suivante

$$w_I = w_1 + w_6$$
$$w_{II} = w_5 \, .$$

A partir des nouveaux coefficients stœchiométriques, on construit les deux réactions globales

$$\text{I:} \qquad 3H_2 + O_2 = 2H + 2H_2O \, , \qquad w_I = w_1 + w_6$$
$$\text{II:} \qquad 2H + M = H_2 + M \, , \qquad w_{II} = w_5$$

Les réactions 1 et 6 représentent la chaîne réactionnelle de ramification, qui forme le radical H, alors que la réaction 5 est la chaîne de rupture. Il est interessant de noter qu'en faisant l'hypothèse de stationnarité pour O et OH, la réaction 1 intervient pour la formation de H (et non pour sa consommation, comme cela peut apparaître dans le mécanisme initial). Cela met en évidence sa propriété de ramification. Les concentrations des radicaux O, OH et HO_2 sont calculées à partir des relations (2.16), (2.17) et (2.21) et des concentrations en H, H_2, O_2 et H_2O. On obtient les relations algébriques non linéaires suivantes

pour OH:
$$\begin{aligned} &2k_{f4}\, C_{OH}^2 + (k_{b1}\, C_O + k_{b2}\, C_H + k_{f3}\, C_{H_2} + k_8\, C_{HO_2})\, C_{OH} \\ &= k_{f1}\, C_H\, C_{O_2} + k_{f2}\, C_O\, C_{H_2} + k_{b3}\, C_{H_2O}\, C_H \\ &\quad + 2k_{b4}\, C_{H_2O}\, C_O + 2k_6\, C_H\, C_{HO_2} \end{aligned} \tag{2.26}$$

pour O:
$$\begin{aligned} &(k_{b1}\, C_{OH} + k_{f2}\, C_{H_2} + k_{b4}\, C_{H_2O})\, C_O \\ &= k_{f1}\, C_H\, C_{O_2} + k_{b2}\, C_{OH}\, C_H + k_{f4}\, C_{OH}^2 \end{aligned} \tag{2.27}$$

pour HO_2:
$$(k_6\, C_H + k_7\, C_H + k_7\, C_H + k_8\, C_{OH})\, C_{HO_2} = k_5\, C_H\, C_{O_2}\, C_M \tag{2.28}$$

qui ne peuvent être résolues que de façon itérative. Toutefois, une simplification importante peut être apportée en négligeant dans la relation (2.26) tous les termes devant $k_{f3}C_{H_2}$ et $k_{b3}C_{H_2O}C_H$. C'est à dire que seule la réaction 3 contribue à la formation et consommation de OH. On peut alors exprimer de façon explicite les concentrations de OH, O et HO_2

$$C_{OH} = \frac{C_H\, C_{H_2O}}{C_{H_2} K_{C_3}} \, , \tag{2.29}$$

$$C_O = \frac{k_{f1}\, C_H\, C_{O_2} + k_{2b}\, C_{OH}\, C_H + k_{f4}\, C_{OH}^2}{k_{1b}\, C_{OH} + k_{2f}\, C_{H_2} + k_{b4}\, C_{H_2O}} \, , \tag{2.30}$$

$$C_{HO_2} = \frac{k_5\, C_H\, C_{O_2}\, C_M}{k_6\, C_H + k_7\, C_H + k_8\, C_{OH}} \, . \tag{2.31}$$

Les vitesses de réaction w_1, w_6 et w_5, qui apparaissent dans (2.22)–(2.25) peuvent alors s'exprimer en fonction des concentrations des espèces non stationnaires H, O_2,

H_2, H_2O. Il est toujours possible de construire un schéma de réactions globales indépendantes les unes des autres: on a quatre espèces non stationnaires, deux éléments O et H fournissent deux relations de conservation (voir éqation (1.7)). Le schéma doit donc nécessairement comprendre au moins $4 - 2 = 2$ réactions globales.

Méthode de réduction à partir des chaînes principales

Les réactions globales peuvent être obtenues en éliminant les espèces stationnaires dans les chaînes principales: La réaction 1 de ramification est à l'origine de la première chaîne principale. Les radicaux O et OH, stationnaires, sont éliminés en ajoutant les réactions 2 et 3, qui sont des réactions très rapides de consommation de O et OH

1	$H + O_2 = OH + O$
2	$O + H_2 = OH + H$
2×3	$2OH + 2H_2 = 2H_2O + 2H$
I	$O_2 + 3H_2 = 2H_2O + 2H$

La réaction 3 étant très rapide, celle-ci peut intervenir plusieurs fois dans la chaîne principale. La deuxième chaîne principale est la chaîne de rupture. Celle-ci comporte les réactions 5 et 7, qui permettent d'éliminer HO_2

5	$H + O_2 + M = HO_2 + M$
7	$H + HO_2 = H_2 + O_2$
II	$H + H + M = H_2 + M$

Les deux réactions globales étant ainsi établies, il s'agit d'étudier l'influence des réactions qui n'ont pas été prises en compte.

Influence de la réaction 6:

6	$H + HO_2 \rightarrow OH + OH$
-7	$H_2 + O_2 \rightarrow H + HO_2$
2×3	$2OH + 2H_2 \rightarrow 2H_2O + 2H$
	$3H_2 + O_2 \rightarrow 2H + 2H_2O$

La réaction 6 a la même influence que la réaction globale I, à savoir production des radicaux H. Sa vitesse de réaction w_6 apparaîtra dans la vitesse globale w_I.

Influence de la réaction 8:

8	$OH + HO_2 \rightarrow H_2O + O_2$
−7	$H_2 + O_2 \rightarrow HO_2 + H$
	$OH + H_2 \rightarrow H_2O + H$

La réaction 8 est identique à la réaction 3, qui a été éliminée dans la chaîne principale I. Celle-ci peut donc aussi être éliminée.

En procédant de la même manière avec toutes les réactions élémentaires, on obtient les vitesses de réaction globales

$$w_I = w_1 + w_6 ,$$
$$w_{II} = w_5 .$$

Mécanisme réduit pour le méthane

Les espèces, dont la concentration maximale reste très faible lors de la combustion du méthane, sont, en plus de OH, O, HO_2, les espèces intermédiaires CH_3, CH_2O, CHO et CH_3O. Elles sont considérées comme stationnaires. Le radical H, bien que se trouvant en faible concentration, ne doit pas être pris comme espèce stationnaire. En effet, H est une espèce très diffusive, et si l'on pondère sa concentration par son coefficient de diffusion, on remarque que celle-ci devient importante. Le terme $L(C_H)$ dans (2.14) ne peut être négligé. Pour la combustion du méthane, quatre réactions globales sont nécessaires. La chaîne principale de consommation du méthane est

11	$CH_4 + H = CH_3 + H_2$
13	$CH_3 + O = CH_2O + H$
14	$CH_2O + H = CHO + H_2$
17	$CHO + M = CO + H + M$
2*b*	$H + OH = O + H_2$
3*b*	$H + H_2O = OH + H_2$
I	$CH_4 + 2H + H_2O = CO + 4H_2$

Cette chaîne décrit l'oxydation du méthane en CO, qui est une espèce non stationnaire. Pour l'oxydation du CO en CO_2, une autre réaction globale est nécessaire

9	$CO + OH = CO_2 + H$
3*b*	$H_2O + H = H_2 + OH$
II	$CO + H_2O = CO_2 + H_2$

Avec les deux réactions globales du système H_2/O_2, on obtient un mécanisme à quatre étapes

I	$CH_4 + 2H + H_2O = CO + 4H_2$
II	$CO + H_2O = CO_2 + H_2$
III	$2H + M = H_2 + M$
IV	$3H_2 + O_2 = 2H + 2H_2O$

De la même manière que précédement, on peut déterminer les vitesses de réaction des réactions globales

$$\begin{aligned} w_I &= w_{10} + w_{11} + w_{12} \\ w_{II} &= w_9 \\ w_{III} &= w_5 - w_{10} + w_{16} - w_{18} + w_{19} - w_{22} + w_{24} + w_{25} \\ w_{IV} &= w_1 + w_6 + w_{18} + w_{22} \,. \end{aligned}$$

Le même résultat aurait pu être obtenu en prenant comme première réaction de la chaîne 1 la réaction 12. w_{12} doit apparaître dans l'expression de w_I. En utilisant la réaction 10, on obtient la réaction globale

$$CH_4 + H_2O = CO + 3H_2 \,.$$

Celle-ci peut être obtenue par combinaison linéaire de I et III. Cela signifie que w_{10} apparait dans l'expression de w_I, et dans l'expression de w_{III} avec un signe négatif. Nous avons passé en revue toutes les réactions faisant intervenir CH_4. On obtient bien, pour w_I

$$w_I = w_{10} + w_{11} + w_{12} \,.$$

Une chaîne parallèle doit être construite pour éliminer l'espèce stationnaire CH_3O

18	$CH_3 + O_2 \rightarrow CH_3O + O$
20	$CH_3O + M \rightarrow CH_2O + H + M$
−13	$CH_2O + H \rightarrow CH_3 + O$
	$O_2 \rightarrow 2O$

Ce résultat peut être obtenu à partir des réactions globales IV et IIIb

$$\begin{array}{lrcl}
\text{IV} & O_2 + 3\,H_2 & = & 2\,H_2O + 2\,H \\
2\times 3b & 2\,H_2O + 2\,H & = & 2\,OH + 2\,H_2 \\
2\times 2b & 2\,OH + 2\,H & = & 2\,H_2 + 2\,O \\
\text{III}b & 2\,H & = & H_2 \\
\hline
 & O_2 & = & 2\,O
\end{array}$$

Cela signifie que w_{18} doit apparaître dans l'expression de w_{IV}, et dans l'expression de w_{III} avec un signe négatif. Et ainsi de suite. Les vitesses des réactions globales s'expriment en fonction des concentrations des espèces stationnaires OH, O, HO_2, CH_3 et CHO. (L'espèce H_2O_2 peut être négligée pour la combustion du méthane.) Les conditions de stationnarité donnent les équations algébriques suivantes

$$C_{CH_3} = \frac{(k_{11f}\,C_H + k_{12f}\,C_{OH})\,C_{CH_4}}{k_{11b}\,C_{H_2} + k_{12b}\,C_{H_2O} + k_{18}\,C_{O_2} + k_{10b}\,C_H + k_{13}\,C_O},$$

$$C_{CHO} = \frac{(k_{13f}\,C_O + k_{18}\,C_{O_2})\,C_{CH_3}}{k_{16}\,C_H + k_{17}\,C_M}.$$

Pour les radicaux O, OH et HO_2, on peut avec une bonne approximation réutiliser les expressions (2.29), (2.30), (2.31) obtenues précédement.

3. Explosion Dans Les Systèmes Homogènes

Dans ce chapitre, nous aborderons la combustion en fonction du temps d'un système homogène, c'est à dire pour lequel les phénomènes de transport (diffusion conduction) n'interviennent pas. Lorsque la température du système augmente, la vitesse des réactions cinétiques devient plus importante, jusqu'à explosion du mélange. La théorie de l'explosion a été developpée par les chercheurs russes, comme Semjonov, Frank-Kamenetzkii, Zeldovich et d'autres.

Equations de base pour la combustion homogène à volume constant

Soit un récipient fermé, contenant un mélange explosif. Si la température du système augmente, il arrive un état où le mélange s'enflamme. On suppose, juste avant l'explosion, que la pression, la température et la densité restent constantes, et que le système est au repos. Les parois de l'enceinte sont prises adiabatiques, et pour éviter les réactions hétérogènes, non catalytiques. Pour un tel système fermé, à volume constant, la densité est constante

$$\mathrm{d}\rho = -\rho^2 \mathrm{d}v = 0\,. \tag{3.1}$$

D'après le premier principe de la thermodynamique (1.25), on a, avec $\mathrm{d}q = 0$ et $\mathrm{d}w_R = 0$

$$\frac{\mathrm{d}u}{\mathrm{d}t} = 0\,. \tag{3.2}$$

L'équation de conservation des espèces, à ρ constant, s'écrit

$$\rho\,\frac{\partial Y_i}{\partial t} = M_i \sum_{k=i}^{r} \nu_{ik}\, w_k = \dot{m}_i \qquad (i = 1, 2, ..., n)\,. \tag{3.3}$$

Par définition de l'énergie interne (1.19) on a

$$\mathrm{d}u = \sum_{i=1}^{n} u_i\,\mathrm{d}Y_i + \sum_{i=1}^{n} Y_i\,\mathrm{d}u_i\,. \tag{3.4}$$

Le dernier terme pouvant s'écrire

$$\sum_{i=1}^{n} Y_i\,\mathrm{d}u_i = \sum_{i=1}^{n} Y_i\,c_{vi}\,\mathrm{d}T = c_v\,\mathrm{d}T\,. \tag{3.5}$$

Des relations (3.3) et (3.5) on tire

$$\rho c_v\,\frac{\mathrm{d}T}{\mathrm{d}t} = -\sum_{i=1}^{n} u_i\,M_i \sum_{k=1}^{r} \nu_{ik}\,w_k = -\sum_{i=1}^{n} u_i\,\dot{m}_i = \sum_{k=1}^{r}(-\Delta U_k)\,w_k\,. \tag{3.6}$$

Théorie de l'explosion thermique adiabatique

Nous prendrons l'hypothèse la plus simple, d'une réaction globale unique, entre le combustible et l'oxydant

$$\nu_F \mathrm{F} + \nu_{O_2} \mathrm{O}_2 \quad \longrightarrow \quad \text{Produits}$$

de vitesse de réaction

$$w = B\left(\frac{\rho Y_F}{M_F}\right)\left(\frac{\rho Y_{O_2}}{M_{O_2}}\right)\exp\left(-\frac{E}{RT}\right). \tag{3.7}$$

Les équations de conservation des espèces et de l'énergie prennent la forme

$$\rho \frac{\mathrm{d}Y_i}{\mathrm{d}t} = M_i \nu_i w\,, \tag{3.8}$$

$$\rho c_v \frac{\mathrm{d}T}{\mathrm{d}t} = (-\Delta U)\, w\,. \tag{3.9}$$

En supposant que c_v et $(-\Delta U)$ restent constants, on obtient la relation de couplage

$$\rho \frac{\mathrm{d}}{\mathrm{d}t}\left(\frac{c_v T}{-\Delta U}\right) = \rho \frac{\mathrm{d}}{\mathrm{d}t}\left(\frac{Y_i}{M_i \nu_i}\right) = w \tag{3.10}$$

qui peut être intégrée entre l'instant t et l'instant initial $t = 0$, pour lequel $Y_i = Y_{i0}$ et $T = T_0$

$$\frac{c_v (T - T_0)}{(-\Delta U)} = \frac{Y_i - Y_{i0}}{\nu_i M_i}\,. \tag{3.11}$$

Nous ne nous interessons qu'à de petites variations de température et de concentration. On peut donc écrire un développement limité de ces valeurs autour de l'état initial

$$\begin{aligned} T &= T_0 + \varepsilon T_1\,, \\ Y_i &= Y_{i0} + \varepsilon Y_{i1}\,. \end{aligned} \tag{3.12}$$

ε étant un paramètre petit, que nous chercherons à déterminer par la suite. En effectuant un développement en série de $1/T$

$$\frac{1}{T} = \frac{1}{T_0} - \frac{(T - T_0)}{T_0^2} = \frac{1}{T_0} - \frac{\varepsilon T_1}{T_0^2} \tag{3.13}$$

le terme exponentiel de la vitesse de réaction s'écrit

$$\exp\left(-\frac{E}{RT}\right) = \exp\left(-\frac{E}{RT_0}\right)\exp\left(\frac{E}{RT_0^2}\,\varepsilon T_1\right). \tag{3.14}$$

Une définition peut-être prise pour ε

$$\varepsilon = \frac{R T_0^2}{E}, \tag{3.15}$$

qui permet d'obtenir l'expression la plus simple pour w. ε étant une grandeur très petite, cette définition n'est valable que si l'énergie d'activation de la réaction globale unique est très grande. En introduisant (3.14) dans l'équation de conservation de l'énergie (3.9) on obtient

$$c_v \, \varepsilon \, \frac{\mathrm{d}T_1}{\mathrm{d}t} = (-\Delta U) \, B \, \rho \left(\frac{Y_{F0} + \varepsilon \, Y_{F1}}{M_F} \right) \left(\frac{Y_{O_20} + \varepsilon \, Y_{O_21}}{M_{O_2}} \right) \exp\left(-\frac{E}{R T_0} \right) \exp T_1 \, . \tag{3.16}$$

Les termes εY_{F1} et εY_{O_21} peuvent être négligés. Cela conduit à l'expression simplifiée

$$\frac{\mathrm{d}T_1}{\mathrm{d}t} = \frac{(-\Delta U) \, B \, \rho}{\varepsilon c_v} \left(\frac{Y_{F0}}{M_F} \right) \left(\frac{Y_{O_20}}{M_{O_2}} \right) \exp\left(-\frac{E}{R T_0} \right) \exp T_1 \, . \tag{3.17}$$

On peut définir un temps caractéristique t_i

$$t_i = \frac{R T_0^2}{E} \, \frac{c_v}{(-\Delta U) \, B \, \rho} \left(\frac{M_F}{Y_{F0}} \right) \left(\frac{M_{O_2}}{Y_{O_20}} \right) \exp\left(\frac{E}{R T_0} \right), \tag{3.18}$$

de telle sorte que l'équation pour la température se mette sous la forme

$$\frac{\mathrm{d}T_1}{\mathrm{d}(t/t_i)} = \exp T_1 \tag{3.19}$$

dont la solution est

$$T_1 = -\ln\left(1 - \frac{t}{t_i} \right) . \tag{3.20}$$

Cela indique que la température augmente très fortement lorsque le temps t s'approche du temps caractéristique t_i. Le temps t_i représente ainsi le délai d'auto-inflammation. L'évolution de la température en fonction du temps est représentée figure 3.

Effet des pertes de chaleur

Pour tenir compte de l'effet des pertes de chaleur, un terme d'échange de chaleur doit être rajouté à l'équation de conservation de l'énergie

$$\rho \, c_v \, \frac{\partial T}{\partial t} = (-\Delta U) \, w - h \, (T - T_0) \tag{3.21}$$

où $h = \rho c_v / t_q$, t_q est un temps caractéristique des pertes de chaleur. Les deux termes du membre de droite sont représentés en fonction de la température sur la figure 4.

Suivant la valeur que prend h, trois cas peuvent se présenter:

a) aucun point d'intersection,
b) les deux courbes se croisent en deux points,
c) les deux courbes se croisent en un point.

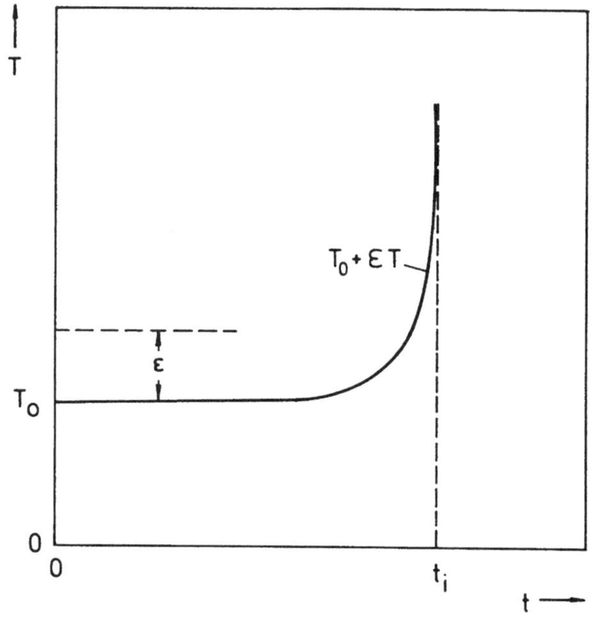

Figure 3:

Evolution de la température en fonction du temps, pour un système homogène fermé, à volume constant.

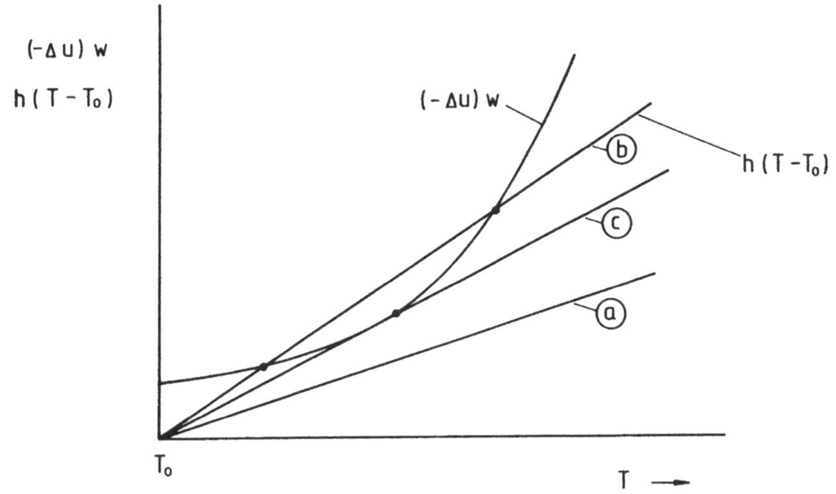

Figure 4:

Comparaison entre les termes production de chaleur et perte de chaleur en fonction de la température, pour un mélange homogène, à volume constant.

Dans le premier cas, le terme de réaction est toujours supérieur au terme de perte de chaleur. Ce cas est comparable à celui traité ci-dessus: Une explosion apparait lorsque le délai d'auto-inflammation est atteint. Dans le cas b) la température augmente jusqu'à ce qu'elle ait atteint le premier point d'intersection. Dans le cas où la température continuerait à augmenter, les pertes de chaleur deviennent plus importantes que la chaleur dégagée par la réaction, et la température décroit. Si on ne tient pas compte de la consommation du mélange initial, on se trouve à un point stable qui représente une solution stationnaire, avec une température constante. Aucune explosion ne peut avoir lieu. Le cas a) est le cas limite: Les pertes de chaleur ne peuvent en aucun cas être supérieur à la chaleur libérée par la réaction. Une solution stationnaire ne peut se présenter qu'après un temps infiniement long. Cela signifie aussi que le délai d'auto-inflammation est infini.

En reprenant le développement asymptotique à partir de (3.11), on obtient

$$\frac{\mathrm{d}T_1}{\mathrm{d}(t/t_i)} = \exp T_1 - \alpha\, T_1 \tag{3.22}$$

avec

$$\alpha = \frac{t_i}{t_q}\,.$$

Le délai d'auto-inflammation $t_{i,q}$ est déterminé à partir du calcul de l'intégrale

$$\frac{t_{i,q}}{t_i} = \int_0^{\infty} \frac{\mathrm{d}T_1}{\exp T_1 - \alpha\, T_1}\,. \tag{3.23}$$

Lorsque le cas b) tend vers le cas c), on a égalité entre le terme de réaction et les pertes de chaleur, et égalité entre leur dérivée respectives

$$\exp T_1 = \alpha\, T_1 \tag*{(3.24)$_1$}$$

et

$$\frac{\mathrm{d}(\exp T_1)}{\mathrm{d}T_1} = \alpha\,. \tag*{(3.24)$_2$}$$

La solution est $T_1 = 1$ et $\alpha = e$. A proximité de ce point, on peut écrire

$$\alpha = (1-\varepsilon)\, e\,, \qquad T_1 = 1 - \varepsilon\, y\,, \tag{3.25}$$

ε petit. Le signe négatif signifie que l'on se trouve en dessous de la courbe correspondant au cas a). En remplaçant (3.25) dans (3.23) on obtient

$$\begin{aligned} \frac{t_{i,q}}{t_i} &= \lim_{\varepsilon \to 0} \int_{1/\varepsilon}^{-\infty} \frac{-\varepsilon\, \mathrm{d}y}{e\,(1 - \varepsilon y + \varepsilon^2 y^2/2) - e\,(1-\varepsilon)(1-\varepsilon y)} \\ &= \frac{1}{e}\sqrt{\frac{2}{\varepsilon}} \int_{-\infty}^{+\infty} \frac{\mathrm{d}x}{1+x^2} = \pi \sqrt{\frac{2}{e\,(e-\alpha)}} \end{aligned} \tag{3.26}$$

avec $x^2 = \varepsilon y^2/2$. Pour $\alpha = e$, le délai d'auto-inflammation, devient infiniement grand.

4. Le Réacteur Parfaitement Mélangé

Prenons le cas d'un réacteur dans lequel les mouvements turbulents sont suffisamment importants pour supposer que les concentrations et la température sont homogènes. Cela signifie qu'à son admission, le mélange se trouve immédiatement à l'état régnant dans le réacteur: T et Y_i.

Equations décrivant l'évolution du mélange en fonction du temps

Le réacteur a un volume V, et contient une masse de gaz m. Il est traversé par un flux massique $\dot{m}$ qui se trouve à l'origine à la température T_u, dont la composition est donnée par les fractions massiques $Y_{i,u}$. A la sortie du réacteur, le mélange est à T_b, les concentrations massiques notées $Y_{i,b}$. La figure 5 montre une coupe schématique d'un réacteur de "Longwell et Weiss". La modification au cours du temps de la masse du composant i est donnée par

$$\frac{\mathrm{d}m_i}{\mathrm{d}t} = (\dot{m}_u - \dot{m}_b)_i + \left(\frac{\mathrm{d}m_i}{\mathrm{d}t}\right)_{\text{chem}} . \tag{4.1}$$

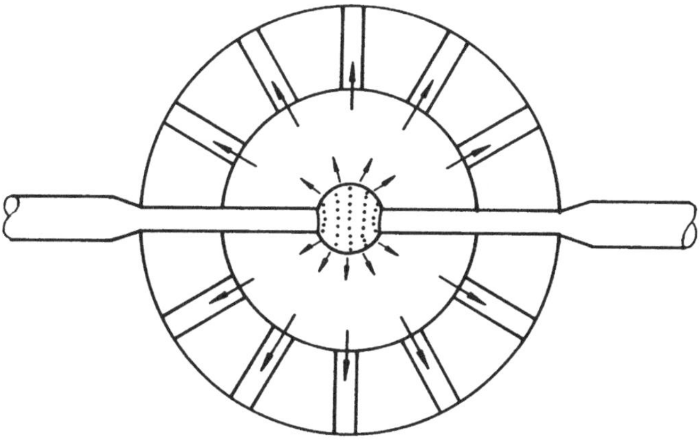

Figure 5:

Coupe schématique d'un réacteur de "Longwell et Weiss".

Les flux massiques à l'entrée et à la sortie sont proportionnels aux fractions massiques respectives

$$(\dot{m}_u - \dot{m}_b)_i = \dot{m}(Y_{i,u} - Y_{i,b}) = \dot{m}(Y_{i,u} - Y_i) \,. \tag{4.2}$$

La variation de la masse de i sous l'action des réactions chimiques peut être écrite comme étant le produit du volume V et la variation de densité (due aux réactions chimiques)

$$\left(\frac{\mathrm{d}m_i}{\mathrm{d}t}\right)_{\mathrm{chem}} = V\dot{m}_i \,. \tag{4.3}$$

Avec $m_i = Y_i m$, m étant invariable au cours du temps, on a

$$m\,\frac{\mathrm{d}Y_i}{\mathrm{d}t} = \dot{m}(Y_{i,u} - Y_i) + V\dot{m}_i \,. \tag{4.4}$$

Le réacteur est supposé adiabatique. L'enthalpie reste donc constante

$$\mathrm{d}h = \sum_{i=1}^{n} h_i\,\mathrm{d}Y_i + \sum_{i=1}^{n} Y_i\,\mathrm{d}h_i = 0\,. \tag{4.5}$$

Avec

$$\sum_{i=1}^{n} Y_i\,\mathrm{d}h_i = c_p\,\mathrm{d}T \tag{4.5}$$

et (4.4) on obtient l'équation suivante pour la température

$$m\,c_p\,\frac{\mathrm{d}T}{\mathrm{d}t} = \dot{m}\,c_p(T_u - T) - V\sum_{i=1}^{n} h_i\,\dot{m}_i \,. \tag{4.6}$$

Prenons le cas simple d'une réaction globale unique, et d'un mélange soit très pauvre, soit très riche, de telle sorte que la mélange dans le réacteur est constitué uniquement soit d'oxygène, soit de combustible. La vitesse de réaction s'écrit

$$\dot{m}_i = \nu_i\,M_i\,w = -B\,\rho\,Y \exp\left(-\frac{E}{RT}\right) . \tag{4.7}$$

$\nu_F = -1$ dans le cas d'un mélange initial très pauvre, $\nu_O = -1$ dans le cas d'un mélange initial très riche. Y représente la fraction massique de l'espèce qui fait défault $1 - Y \simeq 1$. D'après la conservation de la masse avant et après passage dans le réacteur, le temps de résidence du gaz dans celui-ci est

$$t_v = \frac{m}{\dot{m}}\,, \tag{4.8}$$

qui représente le temps caractéristique du mélange. En divisant (4.4) et (4.6) par m, on obtient les équations instationnaires pour les conditions d'un réacteur parfaitement mélangé

$$\frac{\mathrm{d}Y}{\mathrm{d}t} = \frac{Y_u - Y}{t_v} - B\,Y \exp\left(-\frac{E}{RT}\right), \qquad (4.9)$$
$$\frac{\mathrm{d}T}{\mathrm{d}t} = \frac{T_u - T}{t_v} + \frac{(-\Delta H)}{c_p\,M}\,B\,Y \exp\left(-\frac{E}{RT}\right).$$

Ici, M représente la masse molaire de l'espèce faisant défault. On supposera dans ce qui suit que $(-\Delta H)$ et c_p sont constants. Il est pratique d'utiliser des grandeurs adimensionelles, en rapportant les inconnues Y et T aux valeurs initiales T_u et Y_u. Le temps est rapporté au temps de résidence

$$Y^* = \frac{Y}{Y_u}, \qquad T^* = \frac{T}{T_u}, \qquad t^* = \frac{t\,\dot{m}}{m} = \frac{t}{t_v}, \qquad (4.10)$$
$$\mathrm{Da} = B\,t_v, \qquad E^* = \frac{E}{R\,T_u}, \qquad Q^* = \frac{(-\Delta H)\,Y_u}{c_p\,M\,T_u}.$$

Le nombre de Damköhler, Da est le rapport entre le temps de séjour des gaz dans le réacteur et le temps caractéristique de la réaction chimique $t_r = 1/B$. Dans la suite, les numéros se rapportant à des équations adimensionalisées seront suivies d'une étoile $*$. Les équations instationnaires s'écrivent

$$\frac{\mathrm{d}Y}{\mathrm{d}t} = 1 - Y - \mathrm{Da}\,Y \exp\left(-\frac{E}{T}\right) \qquad (4.11)^*$$
$$\frac{\mathrm{d}T}{\mathrm{d}t} = 1 - T + Q\,\mathrm{Da}\,Y \exp\left(-\frac{E}{T}\right).$$

Solution stationnaire

Pour une solution stationnaire, le système $(4.11)^*$ devient

$$0 = 1 - Y_s - \mathrm{Da}\,Y_s \exp\left(-\frac{E}{T_s}\right), \qquad (4.12)^*$$

$$0 = 1 - T_s + Q\,\mathrm{Da}\,Y_s \exp\left(-\frac{E}{T_s}\right). \qquad (4.13)^*$$

Multiplions $(4.12)^*$ par Q^* et additionnons les deux équations

$$0 = 1 - T_s + Q\,(1 - Y_s). \qquad (4.14)^*$$

Y_s^* peut alors être éliminée dans (4.13)*

$$0 = 1 - T_s + \mathrm{Da}(1 - T_s + Q) \exp\left(-\frac{E}{T_s}\right). \qquad (4.15)^*$$

La relation (4.14)* indique également le domaine dans lequel peut varier T_s

$$\begin{aligned} T_{s,\max} &= T_b = 1 + Q \qquad \text{pour} \quad Y_s = 0, \\ T_{s,\min} &= T_u = 1 \qquad \text{pour} \quad Y_s = 1. \end{aligned} \qquad (4.16)^*$$

Sur la figure 6 sont représentées la courbes T_s^* en fonction du nombre de Damköhler, pour différentes valeurs du paramètre E^*. La chaleur de réaction Q^* a été prise égale à 4, ce qui est une valeur typique pour la combustion.

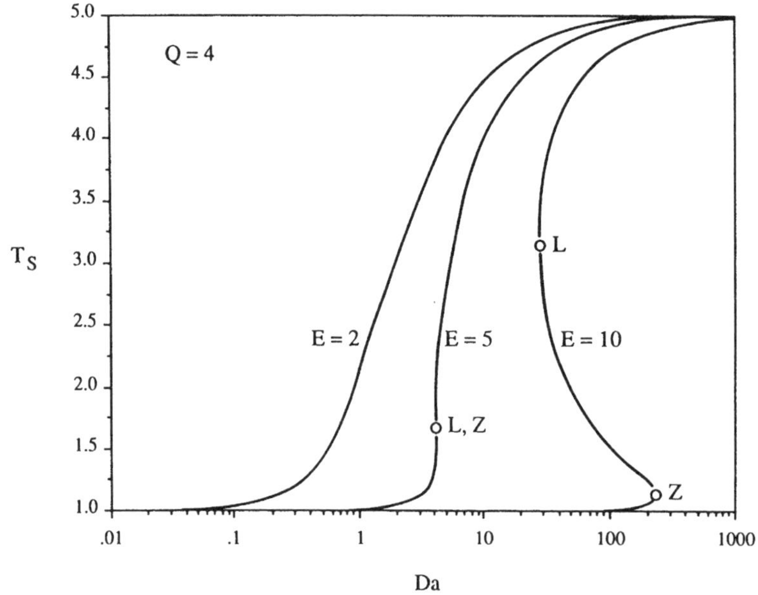

Figure 6:

Evolution de la température en fonction du nombre de Damk"ohler dans un réacteur parfaitement mélangé, et pour des conditions de stationnarité.

Allumage et extinction

Le point d'allumage I et le point d'extinction Q sont caractérisés par la condition

$$\frac{\mathrm{dDa}}{\mathrm{d}T_s} = 0\,. \tag{4.17)*$$

On peut à partir de (4.15)* exprimer le nombre de Damköhler en fonction de la température. Soit la fonction

$$N(T) = (1+Q-T)\exp\left(-\frac{E}{T}\right) \tag{4.18)*$$

on a

$$\mathrm{Da} = \frac{T_s - 1}{N_s}$$

et la pente

$$\frac{\mathrm{dDa}}{\mathrm{d}T_s} = \frac{N_s - (T_s-1)\,\mathrm{d}N_s/\mathrm{d}T_s}{N_s^2}\,. \tag{4.19)*$$

Les conditions d'allumage et d'extinction sont données par

$$N_s - (T_s-1)\,\frac{\mathrm{d}N_s}{\mathrm{d}T_{s,Q,I}} = 0\,. \tag{4.20)*$$

Avec

$$\frac{\mathrm{d}N_s}{\mathrm{d}T_s} = \left(-\frac{1}{1+Q-T_s} + \frac{E}{T_s^2}\right) N_s$$

on obtient finalement

$$\frac{1}{T_{s,Q,I}-1} + \frac{1}{1+Q-T_{s,Q,I}} = \frac{E}{T_{s,Q,I}^2}\,, \tag{4.21)*$$

qui est une équation quadratique pour T_s, dont la solution est

$$T_{s,Q,I} = \frac{E(2+Q) \pm EQ\sqrt{1-4(1+Q)/EQ}}{2(E+Q)}\,. \tag{4.22)*$$

La température d'extinction $T_{s,Q}$ (signe positif) et la température d'allumage $T_{s,I}$ (signe négatif) sont ainsi déterminés. Dans le cas limite où les deux points d'allumage et d'extinction sont confondus, on a la condition

$$E = \frac{4(1+Q)}{Q}\,,$$

ce qui donne pour $Q = 4$ la valeur $E = 5$. Pour une énergie d'activation inférieure, il n'y a plus d'allumage possible. Une courbe en S présentant un point d'allumage et un point d'extinction distincts n'existe que pour

$$E > \frac{4(1+Q)}{Q}.$$

Pour E très grand, on peut développer en série de Taylor les termes suivants

$$\begin{aligned} \sqrt{1 - \frac{4(1+Q)}{EQ}} &= 1 - 2(1+Q)/EQ\,, \\ \left(1 + \frac{Q}{E}\right)^{-1} &= 1 - \frac{Q}{E}\,. \end{aligned} \qquad (4.23)^*$$

En remplaçant $(4.23)^*$ dans $(4.22)^*$ on obtient pour la température d'auto-inflammation, en négligeant les termes en $1/E^2$

$$T_{s,I} = 1 + \frac{1}{E}\,, \qquad (4.24)^*$$

qui s'écrit en fonction des variables réelles

$$T_{s,I} = T_u + \frac{RT_u^2}{E}\,. \qquad (4.25)$$

On voit ainsi que la température d'allumage est très légèrement supériere à la température T_u, le terme RT_u^2/E étant très petit devant T_u. De la même manière, pour la température d'extinction on obtient

$$T_{s,Q} = 1 + Q - \frac{(1+Q)^2}{E}\,, \qquad (4.26)^*$$

qui devient

$$T_{s,Q} = T_b - \frac{RT_b^2}{E}\,. \qquad (4.27)$$

La température d'extinction est très légèrement inférieure à la température maximale T_b.

Limites asymptotiques pour les solutions stationnaires

Pour des très petits nombres de Damköhler posons

$$\mathrm{Da} = \varepsilon\,. \qquad (4.28)$$

La température d'extinction peut être développée autour de T_u

$$T_s = T_u + \varepsilon\, y\,, \tag{4.29)*}$$

de même que la fonction $N(T)$

$$N(T_s) = N(T_u) + \varepsilon \left(\frac{\partial N}{\partial \varepsilon}\right)_{\varepsilon=0}. \tag{4.30)*}$$

On a la relation suivante

$$\left.\frac{\partial N}{\partial \varepsilon}\right|_{\varepsilon=0} = y \times \left.\frac{\partial N}{\partial T}\right|_{T=T_u} = y \times \left(\frac{(1+Q-T_u)\,E}{T_u^2} - 1\right)\exp\left(-\frac{E}{T_u}\right). \tag{4.31)*}$$

A l'aide de (4.28), (4.29)*, (4.30)* et (4.31)*, la relation (4.15)* devient

$$1 - T_u - \varepsilon\, y + \varepsilon\left(N(T_u) + \varepsilon\, y \left(\frac{\partial N}{\partial T}\right)_{T=T_u}\right) = 0\,. \tag{4.32)*}$$

Avec $T_u^* = 1$

$$y \cong N(T_u) = Q\exp(-E)\,, \tag{4.33)*}$$

ainsi, pour de petits nombres de Damköhler

$$T_s = 1 + \mathrm{Da}\, Q\exp(-E) + O(\mathrm{Da}^2)\,. \tag{4.34)*}$$

Cette fonction, monotomme croissante ne prédit pas de point d'allumage. On peut procéder de la même manière pour les grandes nombres de Damköhler. On procède alors à un développement limité de la température autour de T_b

$$T_s = T_b - \varepsilon\, y\,, \qquad \varepsilon = \frac{1}{\mathrm{Da}}\,.$$

La relation (4.32)* prend la forme

$$\varepsilon\,(1 - T_b + \varepsilon\, y) + N(T_b) - \varepsilon\, y \left(\frac{\partial N}{\partial T}\right)_{T=T_b} = 0 \tag{4.35)*}$$

avec $T_b^* = 1 + Q$

$$N(T_b) = 0$$

$$\left(\frac{\partial N}{\partial T}\right)_{T=T_b} = -\exp\left(\frac{-E}{1+Q}\right).$$

Finalement

$$y = Q\exp\left(\frac{E}{1+Q}\right) \tag{4.36)*}$$

et l'expression pour les grands nombres de Damköhler est

$$T_s = 1 + Q - \frac{1}{\text{Da}} Q \exp\left(\frac{E}{1+Q}\right) + O(\frac{1}{\text{Da}^2}). \qquad (4.37)^*$$

Les courbes obtenues par les deux cas étudiés sont présentées sur figure 7 pour $E = 2$ et $E = 10$ respectivement. Dans le cas $E = 2$, le développement asymptotique approche la solution exacte de manière satisfaisante. Par contre, pour $E = 10$, l'écart est important. En fait l'accord est bon tant qu'on se trouve assez loin des points d'allumage et d'extinction, mais les solutions asymptotiques ne prévoient pas leur existence. Prenons le cas d'un grand nombre de Damköhler et d'une grande énergie d'activation, afin d'étudier le comportement de la courbe à proximité du point d'extinction. Reprenons le développement

$$T_s = T_b - \varepsilon y,$$

ε étant cette fois-ci inconnu. En remplaçant cette expression dans $(4.15)^*$ on obtient, avec $1/T_s = 1/T_b - \varepsilon y/T_b^2$

$$-Q + \varepsilon y + \text{Da}\, \varepsilon y \exp\left(-\frac{E}{1+Q}\right) \exp\left(-\frac{E\varepsilon}{(1+Q)^2} y\right) = 0. \qquad (4.38)^*$$

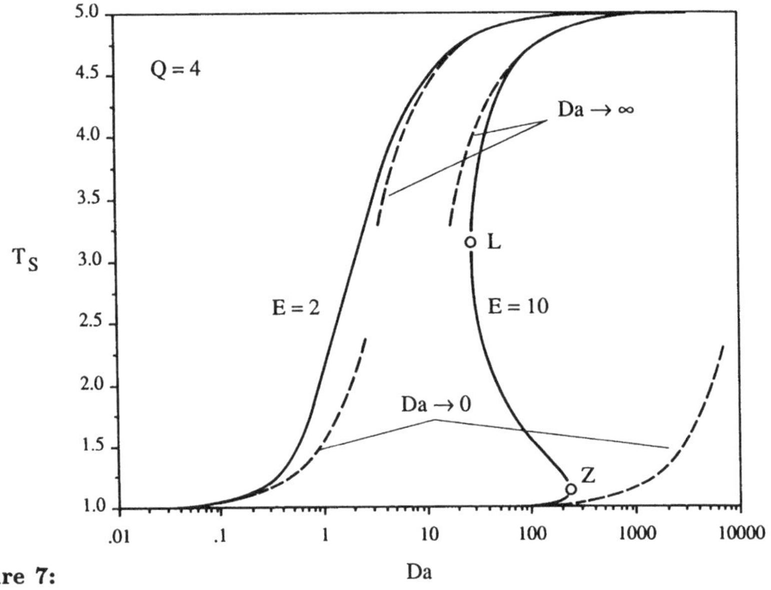

Figure 7:

Evolution de la température en fonction du nombre de Damk"ohler. Les courbes en pointillés sont obtenues à partir de l'étude asymptotique, avec $\varepsilon = 1/\mathrm{Da}$.

Le deuxième terme peut être négligé devant Q qui est de l'ordre de 1. On a alors deux possibilités pour déterminer ε: la prémière $\varepsilon = 1/\mathrm{Da}$ n'est pas pratique, cas alors pour E très grand, $\exp(-E/(1+Q))$ devient infiniment petit et le deuxième terme disparait complètement. La deuxième, utilisée ici, est de choisir ε de telle sorte que la combinaison des paramètres qui apparaissent dans la dernière exponentielle soit égale à 1

$$\varepsilon = \frac{(1+Q)^2}{E}. \tag{4.39}^*$$

Afin que le deuxième terme ne disparaisse pas, il est nécessaire d'avoir la condition

$$\delta = \frac{\mathrm{Da}\,(1+Q)^2}{E\,Q} \exp\left(-\frac{E}{1+Q}\right) \tag{4.40}^*$$

de l'ordre de 1. δ représente un nombre de Damköhler réduit. Cela signifie que la nombre de Damköhler

$$\mathrm{Da} = \delta\,Q\,\varepsilon^{-1} \exp\left(\frac{1+Q}{\varepsilon}\right) \tag{4.41}^*$$

tend vers l'infini, quand ε tend vers 0. y est donc déterminé par la seule relation

$$1 = \delta\,y\,\exp(-y). \tag{4.42}^*$$

Cette expression remplace l'expression (4.36)* lorsque l'énergie d'activation est grande. Pour $y = 1$, $\delta = e$ la fonction présente une tangente verticale qui représente le point d'extinction. En ce point, l'expression de la température est identique à (4.27). On peut procéder de la même manière dans le cas oǹ l'énergie d'activation est très grande, mais le nombre de Damköhler très petit. En introduisant

$$T_s = 1 + \varepsilon\,y$$

dans (4.15)* on obtient

$$-\varepsilon\,y + \mathrm{Da}\,(Q - \varepsilon\,y)\exp(-E)\exp(E\,\varepsilon\,y) = 0, \tag{4.43}^*$$

ce qui suggère comme définission de ε

$$\varepsilon = \frac{1}{E}. \tag{4.44}^*$$

Le nombre de Damköhler réduit prend la forme

$$\Delta = \mathrm{Da}\,Q\,E\,\exp(-E)$$

et la température est déterminée à partir de

$$-y + \Delta \exp y = 0\,,$$

expression qui prévoit également une tangente au point correspondant au point d'allumage. Ces deux cas sont représentés sur les figures 8 et 9.

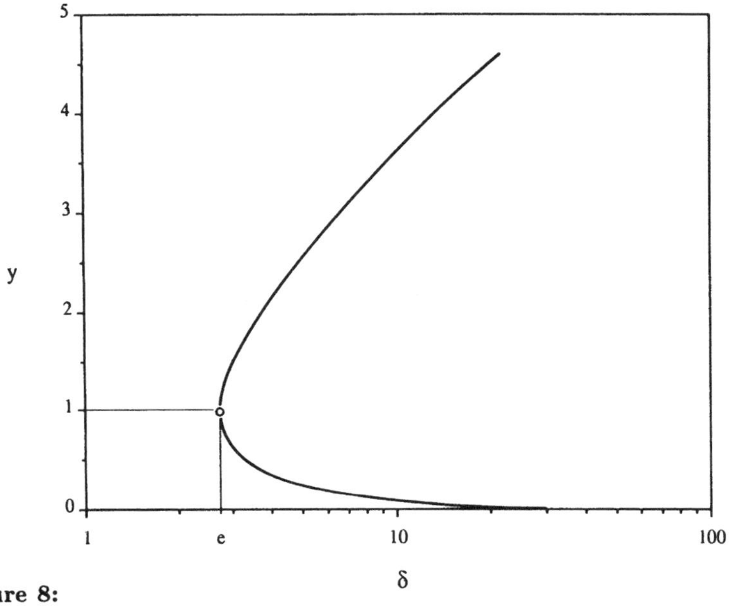

Figure 8:
Evolution de y en fonction du paramètre δ, dans le cas limite d'une grande énergie d'activation, et d'un grand nombre de Damk"ohler.

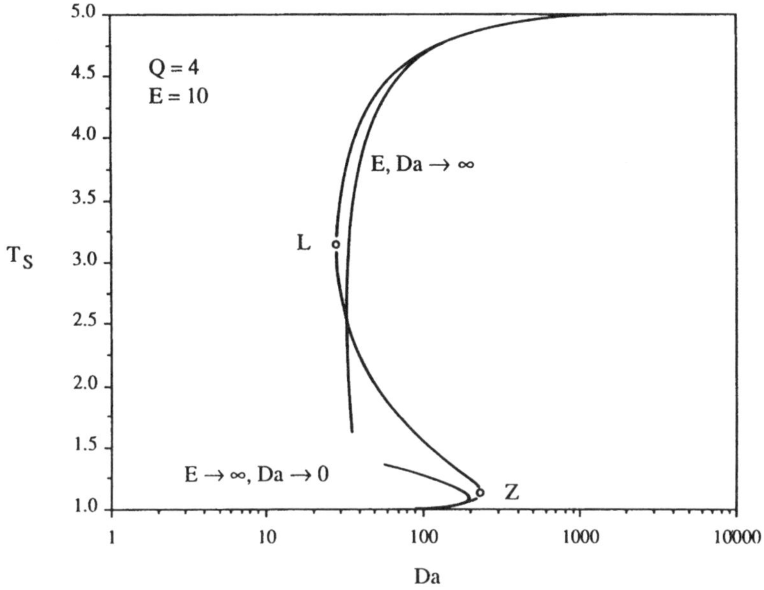

Figure 9:

Evolution de la température en fonction du nombre de Damk"ohler, dans les cas limites où l'énergie d'activation est très grande, et le nombre de Damk"ohler soit très grand, soit très petit.

5. Flammes Laminaires

Les flammes laminaires sont classées en deux catégories: les flammes prémélangées, et les flammes de diffusion. Les prèmieres sont toujours utilisées lorsqu'une combustion intensive doit avoir lieu dans une enceinte petite, comme c'est le cas par exemple dans les moteurs à essence. Les flammes de prémélange ont l'avantage de ne pas produire de suies, et permettent d'atteindre des températures moyennes plus élevées. Le dispositif expérimental caractéristique des flammes prémélangées est le bec-bunsen, dont un schéma de principe est présenté figure 10.

Vitesse de combustion laminaire

Une flamme qui brûle au dessus d'un bec-bunsen prend toujours une forme cônique: celle-ci est stabilisée lorsque la composante normale au front de flamme de la vitesse des gaz frais est égale à la vitesse de combustion. A l'aide d'un bec-bunsen on peut facilement déterminer la vitesse de combustion laminaire s_F

$$s_F = -v_{n,u} = -v_u \sin\alpha\,. \tag{5.1}$$

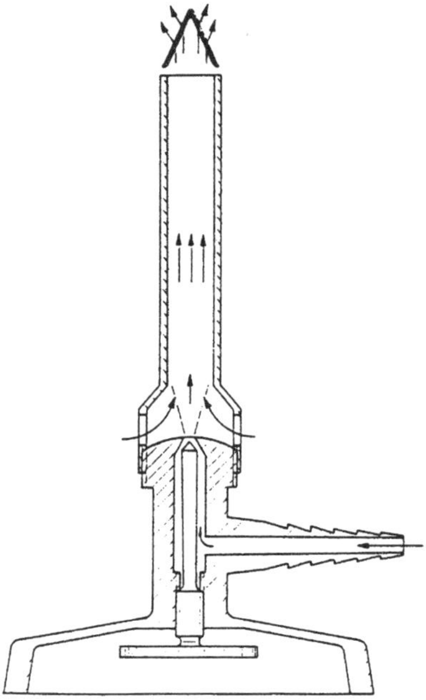

Figure 10:
Schéma de principe d'un bec-bunsen.

v_u étant la vitesse des gaz frais, qui se décompose en une composante tangentielle $v_{t,u}$, et une composante normale $v_{n,u}$, par rapport au front de flamme. α est le demi-angle du "¡cône de Bensen"¿. Une méthode plus exacte pour déterminer s_F consiste à utiliser une chambre de combustion sphérique de volume intérieur suffisament grand, à allumage central. La flamme crée est sphérique, et sa vitesse radiale peut être mesurée. L'élevation de pression dans l'enceinte peut être négligée lorsque le volume inflammable initial de rayon r_F est petit devant le volume total, de rayon R

$$\frac{\Delta p}{p} \simeq \frac{r_F^3}{R^3}\,.$$

La vitesse de combustion laminaire est alors

$$s_{F,b} = \frac{\mathrm{d}r_F}{\mathrm{d}t} \tag{5.2}$$

A partir de l'équation de continuité

$$(\rho v_n)_u = (\rho v_n)_b . \tag{5.3}$$

on obtient la relation suivante

$$\frac{v_{n,b}}{v_{n,u}} = \frac{\rho_u}{\rho_b} \simeq \frac{T_b}{T_u} . \tag{5.4}$$

Au cours de la combustion, la température augmente d'un facteur 4 à 10, la composante normale au front de flamme de la vitesse augmente d'autant, alors que la composante tangentielle reste constante. Dans le cas des flammes instationnaires, la vitesse de propagation de la flamme $\vec{v}_F$ doit être prise en compte: Soit x_F la position du front de flamme et $\vec{n}$ le vecteur normal au front de flamme. On a

$$\begin{aligned} \frac{\mathrm{d}\vec{x}_F}{\mathrm{d}t} \cdot \vec{n} &= \vec{v}_u \vec{n} + s_F , \\ v_{F,n} &= v_{n,u} + s_F . \end{aligned} \tag{5.5}$$

Calcul numérique de la vitesse de combustion

Prenons, pour simplifier les calculs, un système de coordonnées lié à la flamme ($\mathrm{d}x_F/\mathrm{d}t = 0$): L'équation de continuité se met sous la forme

$$\rho v = (\rho v)_u = \rho_u s_F , \tag{5.6}$$

ce qui signifie que le flux massique est constant à travers la flamme. Les équations de conservation des espèces et de la température sont, en négligeant la fonction de dissipation, l'effet Dufour et le rayonnement thermique

$$\begin{aligned} (\rho v)_u c_p \frac{\mathrm{d}T}{\mathrm{d}x} &= \frac{\mathrm{d}}{\mathrm{d}x}\left(\lambda \frac{\mathrm{d}T}{\mathrm{d}x}\right) - \sum_{i=1}^{n} c_{pi} j_i \frac{\mathrm{d}T}{\mathrm{d}x} - \sum_{i=1}^{n} h_i \dot{m}_i , \\ (\rho v)_u \frac{\mathrm{d}Y_i}{\mathrm{d}x} &= -\frac{\mathrm{d}j_i}{\mathrm{d}x} + \dot{m}_i , \end{aligned} \tag{5.7}$$

j_i représentant le flux de diffusion de l'espèce i

$$j_i = \frac{M_i}{M^2} \sum_{\substack{j=1 \\ i \neq j}}^{n} \rho \mathcal{D}_{ij} M_j \frac{\mathrm{d}X_j}{\mathrm{d}x} - \frac{\mathcal{D}_i^T}{T} \frac{\mathrm{d}T}{\mathrm{d}x} . \tag{5.8}$$

Celui-ci s'écrit généralement sous la forme simplifiée

$$j_i = -\rho \mathcal{D}_i \frac{\mathrm{d}Y_i}{\mathrm{d}x}.$$

L'équation de conservation du moment s'écrit simplement

$$P = const.$$

si on suppose que le nombre de Mac est très petit. Au cours de calculs numériques, la valeur du terme $\dot{m}_i$ varie jusqu'à ce qu'une solution stationnaire soit atteinte, qui répond aux conditions limites des gaz frais et des gaz brûlés. Celles-ci sont données par la composition et la température du mélange initial, et la solution d'équilibre dans les gaz brûlés. De tels calculs numériques sont réalisés par Warnatz, par exemple. La figure 11 représente le calcul de vitesses de flamme pour différentes compositions initiales, à partir de mécanismes complet, ou réduit (voir chapitre 2).

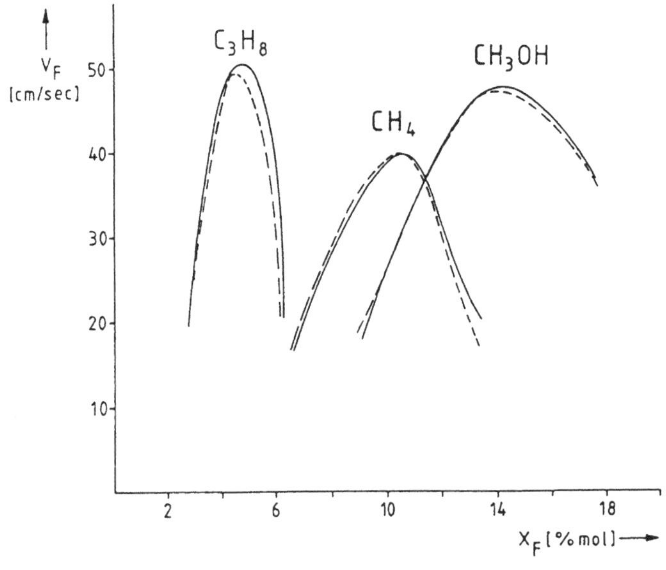

Figure 11:

Calculs de vitesses de flammes à pression atmosphérique, en fonction de la richesse. Les courbes en trait continu représentent les resultats obtenues à partir de mécanismes réduits.

Théorie thermique de la flamme

Le traitement théorique de flammes monodimensionnelles et stationnaires a été réalisé autour de 1938 par Zeldovich et Frank-Kamenetzkii. C'est l'exemple classique d'une description mathématique de la combustion prémélangée. A partir de simplifications systématiques, les équations de base peuvent être écrites sous forme résoluble. On peut voir sur la figure 12 les profils de concentration et de température obtenues pour une flamme pauvre. D'après la théorie de Zeldovich et Frank-Kamenetzkii, la zone de combustion peut être divisée en trois parties: la zone de préchauffage, la zone de réaction et la zone d'équilibre. La première hypothèse est de supposer une réaction globale irréversible à grande énergie d'activation

$$\nu'_F \mathrm{F} + \nu'_{O_2} \mathrm{O_2} \longrightarrow \text{Produits},$$

dont la vitesse est

$$w = B \frac{\rho Y_F}{M_F} \frac{\rho Y_{O_2}}{M_{O_2}} \exp\left(-\frac{E}{RT}\right). \tag{5.9}$$

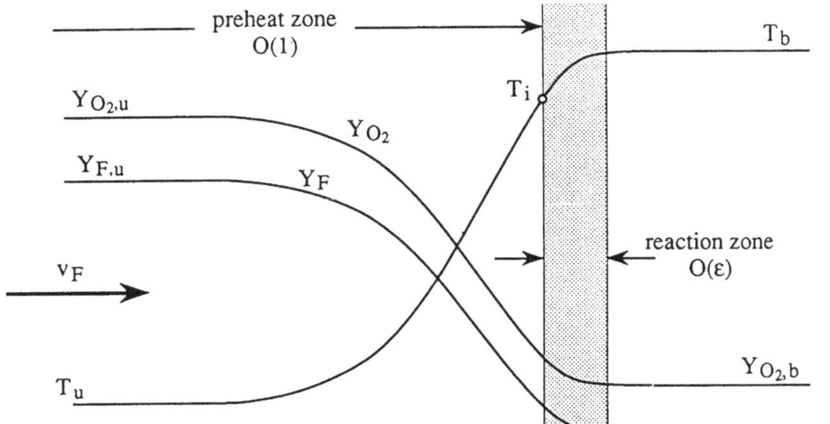

Figure 12:

Repésentation schématique de la structure d'une flamme pauvre prémélangée.

Derrière le front de flamme, un état d'équilibre s'établit, c'est à dire que la réaction chimique ne joue plus aucun rôle $w = 0$. Pour une flamme pauvre le fuel est entièrement consommé, pour une flamme riche, l'oxygène est entièrement consommé, pour une flamme stœchiométrique, Y_F et $Y_{O_2} = 0$. Dans tous les cas, dans cette zone, on a

$$(Y_F Y_{O_2})_b = 0\,. \tag{5.10}$$

Dans la suite, nous ferons l'hypothèse que c_p est constant, identique pour toutes les espèces, de telle sorte que $(-\Delta H)$ soit constant. On supposera de plus que le nombre de Lewis est égal à 1

$$\mathrm{Le}_i = \frac{\lambda}{\rho\, c_p\, \mathcal{D}_i} = 1\,.$$

Les équations de conservation (5.7) se mettent sous la forme simplifiée

$$(\rho\, v)_u \frac{\mathrm{d}T}{\mathrm{d}x} - \frac{\mathrm{d}}{\mathrm{d}x}\left(\frac{\lambda}{c_p}\frac{\mathrm{d}T}{\mathrm{d}x}\right) = \frac{1}{c_p}\, w\,(-\Delta H)\,,$$

$$(\rho\, v)_u \frac{\mathrm{d}Y_i}{\mathrm{d}x} - \frac{\mathrm{d}}{\mathrm{d}x}\left(\frac{\lambda}{c_p}\frac{\mathrm{d}Y_i}{\mathrm{d}x}\right) = M_i\, \nu_i\, w\,.$$

(Le terme convection de chaleur peut être négligé.) En éliminant w, et en intégrant l'équation obtenue, on obtient les relations de couplage entre concentrations et température

$$\begin{aligned} Y_F &= -\frac{\nu'_F\, M_F\, c_p}{(-\Delta H)}\,(T - T_b) + Y_{F,b} \\ Y_{O_2} &= -\frac{\nu'_{O_2} M_{O_2} c_p}{(-\Delta H)}\,(T - T_b) + Y_{O_2,b} \end{aligned} \tag{5.11}$$

L'équation différentielle de conservation de l'énergie permet de déterminer le profil de température suivant la direction x

$$(\rho v)_u \frac{\mathrm{d}T}{\mathrm{d}x} = \frac{\mathrm{d}}{\mathrm{d}x}\left(\frac{\lambda}{c_p}\frac{\mathrm{d}T}{\mathrm{d}x}\right) + \frac{(-\Delta H)}{c_p}\, w(T)\,. \tag{5.12}$$

Les hypothèses de Zeldovich et Frank-Kamenetzkii sont les suivantes:

- dans la zone de préchauffage ($T \le T_i$), aucune réaction cinétique n'a lieu: $w = 0$
- dans la zone de réaction, le terme de convection du membre de gauche de (5.12) peut être négligé devant les termes de diffusion et de réaction.

Dans la zone de préchauffage, l'intégration de l'équation (5.12) donne

$$\frac{\mathrm{d}T}{\mathrm{d}x} = \frac{c_p\,(\rho v)_u}{\lambda}\,(T - T_u)\,. \tag{5.13}$$

Dans la zone de réaction, on a, avec

$$\frac{\mathrm{d}}{\mathrm{d}x}\left(\frac{\lambda}{c_p}\frac{\mathrm{d}T}{\mathrm{d}x}\right) = \frac{\mathrm{d}T}{\mathrm{d}x}\frac{\mathrm{d}}{\mathrm{d}T}\left(\frac{\lambda}{c_p}\frac{\mathrm{d}T}{\mathrm{d}x}\right) \tag{5.14}$$

$$\frac{1}{2}\frac{\mathrm{d}}{\mathrm{d}T}\left(\frac{\lambda}{c_p}\frac{\mathrm{d}T}{\mathrm{d}x}\right)^2 = -\frac{(-\Delta H)\,\lambda}{c_p^2}\,w\,. \tag{5.15}$$

Cette équation peut être intégrée, en utilisant les conditions limites $\mathrm{d}T/\mathrm{d}x = 0$ pour $T = T_b$ et $x \to \infty$

$$\frac{\mathrm{d}T}{\mathrm{d}x} = \sqrt{2\,\frac{(-\Delta H)}{\lambda^2}\int_T^{T_b}\lambda w\,\mathrm{d}T}\,. \tag{5.16}$$

A la frontière des deux zones, pour $T = T_i$, la dérivée de la température est identique

$$\frac{c_p\,(\rho v)_u}{\lambda_i}\,(T_i - T_u) = \sqrt{2\,\frac{(-\Delta H)}{\lambda_i^2}\int_{T_i}^{T_b}\lambda w\,\mathrm{d}T}\,. \tag{5.17}$$

Pour pouvoir interpréter l'intégrale qui apparait dans l'équation, il est nécessaire de faire une hypothèse supplémentaire: T_i peu différent de T_b, et $\rho = \rho_b$, $\lambda = \lambda_b$. Développons alors en série le terme de l'exponentielle dans (5.9)

$$-\frac{E}{RT} = -\frac{E}{R\,T_b} + \frac{E\,(T - T_b)}{R\,T_b^2} \tag{5.18}$$

et posons

$$\Theta = \frac{E\,(T - T_b)}{R\,T_b^2}\,, \tag{5.19}$$

qui représente la nouvelle variable pour la température. Le terme de réaction s'écrit alors, à partir de (5.9) et en utilisant (5.10), (5.11), (5.18) et (5.19)

$$\begin{aligned} w &= B\,\rho_b^2\,\frac{c_p\,R\,T_b^2}{(-\Delta H)\,E}\,\exp\left(-\frac{E}{R\,T_b}\right) \\ &\left[-\left(\frac{\nu_F'\,Y_{O_2,b}}{M_{O_2}} + \frac{\nu_{O_2}'\,Y_{F,b}}{M_F}\right)\Theta + \frac{\nu_{O_2}'\,\nu_F'\,c_p\,R\,T_b^2}{(-\Delta H)\,E}\,\Theta^2\right]\exp\Theta \end{aligned} \tag{5.20}$$

et l'intégrale qui apparait dans (5.17) se met sous la forme

$$\int_{T_i}^{T_b} \lambda w \, \mathrm{d}T = \lambda_b \frac{R T_b^2}{E} \int_{\Theta_i}^{0} w(\Theta) \, \mathrm{d}\Theta$$

$$= \frac{\lambda_b B \rho_b^2 c_p R^2 T_b^4}{(-\Delta H) E^2} \exp\left(-\frac{E}{R T_b}\right) \left[\left(\frac{\nu_F' Y_{O_2,b}}{M_{O_2}} + \frac{\nu_{O_2}' Y_{F,b}}{M_F}\right)\right. \tag{5.21}$$

$$+ \left(1 + (\Theta_i - 1) \exp \Theta_i\right)$$

$$\left. + 2 \frac{\nu_{O_2}' \nu_F' c_p R T_b^2}{(-\Delta H) E} \left(1 - \left(1 - \Theta_i + \frac{\Theta_i^2}{2}\right) \exp \Theta_i\right)\right].$$

Pour une très grande énergie d'activation, on a $\Theta_i \to -\infty$. A partir de (5.17), en posant en plus $T_i = T_b$, on obtient pour la zone de préchauffage la relation suivante pour la vitesse de combustion

$$\rho_u s_F = (\rho v)_u = \sqrt{2 \frac{B \rho_b^2 \lambda_b R^2 T_b^4}{c_p (T_b - T_u)^2 E^2} \exp\left(-\frac{E}{R T_b}\right) S}, \tag{5.22}$$

$$S = \frac{\nu_F' Y_{O_2,b}}{M_{O_2}} + \frac{\nu_{O_2}' Y_{F,b}}{M_F} + 2 \frac{\nu_{O_2}' \nu_F' c_p R T_b^2}{(-\Delta H) E}.$$

Le facteur S dépend fortement de la stœchiométrie du mélange

$$S = \begin{cases} \dfrac{\nu_F' Y_{O_2,b}}{M_{O_2}} & \text{pour} \quad \phi \ll 1 \\ 2 \dfrac{\nu_{O_2}' \nu_F' c_p R T_b^2}{(-\Delta H) E} & \text{pour} \quad \phi = 1 \\ \dfrac{\nu_{O_2}' Y_{F,b}}{M_F} & \text{pour} \quad \phi \gg 1 \end{cases} \tag{5.23}$$

Dérivée de la théorie thermique de la flamme — Développement asymptotique pour de grandes énergies d'activation

Le domaine d'intégration est divisé en trois parties: les zones de préchauffage et d'équilibre sont les zones extérieures entre lesquelles se trouve la zone de réaction. Dans le cas limite où l'énergie d'activation est infiniment grande, l'épaisseur de cette dernière zone est infiniment petite. La structure asymptotique d'une telle flamme est maintenant étudiée. Nous utiliserons pour cela les variables adimensionnelles suivantes

$$x^* = (\rho v)_u \int_0^x \frac{c_p}{\lambda} \, dx$$

$$T^* = \frac{T - T_u}{T_b - T_u}, \qquad T_b - T_u = \begin{cases} \dfrac{(-\Delta H)\, Y_{F,u}}{c_p M_F} & \text{for} \quad \phi \le 1 \\ \dfrac{(-\Delta H)\, Y_{O_2,u}}{c_p\, M_{O_2}} & \text{for} \quad \phi \ge 1 \end{cases} \tag{5.24}$$

$$w^* = \frac{\lambda\,(-\Delta H)\, w}{(\rho\, v)_u^2\, c_p^2\, (T_b - T_u)}$$

L'équation de conservation de l'énergie se met sous la forme

$$\frac{dT}{dx} = \frac{d^2 T}{dx^2} + w \tag{5.25)*$$

avec les conditions aux limites

$$x \to -\infty : \qquad T^* = 0,$$

$$x \to +\infty : \qquad T^* = 1.$$

La zone de réaction est située à $x = 0$. Son épaisseur, de l'ordre de ε, ε étant un paramètre petit inconnu, qui sera déterminé à partir du calcul suivant: Soit la nouvelle coordonnée étendue

$$\zeta = \frac{x}{\varepsilon}. \tag{5.26)*$$

La température T^* est développée autour de 1

$$T = 1 - \varepsilon\, y. \tag{5.27)*$$

Les relations (5.11) deviennent

$$\begin{aligned} Y_F &= Y_{F,b} + \varepsilon\, \frac{c_p\,(T_b - T_u)\, \nu_F'\, M_F}{(-\Delta H)}\, y\,, \\ Y_{O_2} &= Y_{O_2,b} + \varepsilon\, \frac{c_p\,(T_b - T_u)\, \nu_{O_2}'\, M_{O_2}}{(-\Delta H)}\, y\,. \end{aligned} \tag{5.28)*$$

De même, la masse volumique et la conductibilité thermique peuvent se mettre sous la forme

$$\begin{aligned} \rho &= \rho_b - \varepsilon (T_b - T_u) \left.\frac{\partial \rho}{\partial T}\right|_b y\,, \\ \lambda &= \lambda_b - \varepsilon (T_b - T_u) \left.\frac{\partial \lambda}{\partial T}\right|_b y\,. \end{aligned} \tag{5.29}^*$$

A partir du développement du terme exponentiel (5.18) on obtient

$$\begin{aligned} \exp\left(-\frac{E}{RT}\right) &= \exp\left(-\frac{E/R}{T_u + T^*(T_b - T_u)}\right) \\ &= \exp\left(\frac{-E/R}{T_b - (T_b - T_u)\,\varepsilon\, y}\right) \\ &= \exp\left(-\frac{E}{RT_b}\right) \exp\left(-\varepsilon\, y\, \frac{(T_b - T_u)\,E}{RT_b^2}\right). \end{aligned} \tag{5.30}$$

Ceci nous permet de donner une définition pratique pour ε, qui est choisi tel que le deuxième terme de (5.30) se mette sous la forme la plus simple possible soit

$$\varepsilon = \frac{1}{\mathrm{Ze}}\,, \qquad \mathrm{Ze} = \frac{E\,(T_b - T_u)}{RT_b^2}\,. \tag{5.31}$$

Le terme de réaction dans (5.25)* devient, avec $(Y_F Y_{O_2})_b = 0$

$$w = \mathrm{Da}\,\varepsilon\,(y + b\,y^2)\,\exp(-y) \tag{5.32}^*$$

avec

$$\begin{aligned} \mathrm{Da} &= \frac{B\,\rho_b^2\,\lambda_b}{(\rho v)_u^2}\,\frac{\left(Y_{O_2,b}\,\nu_F'\,M_F + Y_{F,b}\,\nu_{O_2}'\,M_{O_2}\right)}{M_F\,M_{O_2}}\,\exp\left(-\frac{E}{RT_b}\right), \\ b &= \frac{\varepsilon\, c_p\,(T_b - T_u)\,\nu_F'\,\nu_{O_2}'\,M_F\,M_{O_2}}{\left(Y_{F,b}\,\nu_{O_2}'\,M_{O_2} + Y_{O_2,b}\,\nu_F'\,M_F\right)(-\Delta H)}\,. \end{aligned} \tag{5.33}$$

En insérant (5.32)* dans (5.25)*, on obtient avec (5.27)*

$$-\varepsilon\,\frac{\mathrm{d}y}{\mathrm{d}\zeta} + \frac{\mathrm{d}^2 y}{\mathrm{d}\zeta^2} = \mathrm{Da}\,\varepsilon^2\,(y + b\,y^2)\,\exp(-y)\,. \tag{5.34}^*$$

Zone de réaction:

On remarque, que dans la zone de réaction, le premier terme du membre de gauche correspondant au terme de convection est très petit, et peut être négligé devant le terme de diffusion. (Dans la zone de réaction $y = O(1)$.) De prime abord, le terme du membre de droite parait également petit. Si ce terme est négligé, cela signifie que dans la zone de réaction aucune réaction n'a lieu, ce qui est une contradiction en soi. Pour que le terme de réaction soit du même ordre de grandeur que le terme de diffusion, il faut que le nombre de Damk"ohler soit de l'ordre de grandeur $O(\varepsilon^{-2})$. Posons

$$\Lambda = \mathrm{Da}\,\varepsilon^2 \simeq O(1)\,. \tag{5.35)*}$$

La relation (5.34)* s'écrit finalement, pour la zone de réaction

$$\frac{\mathrm{d}^2 y}{\mathrm{d}\zeta^2} = \Lambda\,(y + b\,y^2)\exp(-y)\,. \tag{5.36)*}$$

Zone de préchauffage:

En introduisant (5.27)* dans (5.32)* on obtient l'expression suivante pour la vitesse de réaction adimensionnelle

$$w = \mathrm{Da}\left((1-T) + b\,\frac{(1-T)^2}{\varepsilon}\right)\exp\left(-\,\frac{1-T}{\varepsilon}\right)\,. \tag{5.37)*}$$

Dans la zone de préchauffage, $1 - T^*$ est de l'ordre de grandeur $O(1)$, et dans le cas limite, où $\varepsilon \to 0$, la vitesse de réaction devient exponentiellement petite. Le terme de réaction peut donc effectivement être négligé dans cette zone. La solution de l'équation (5.25)* est alors

$$T = \exp x \qquad \text{pour} \qquad x < 0\,, \tag{5.38)*}$$

en tenant compte de la condition limite $T^* = 0$ pour $x^* \to -\infty$. Pour $x^* = 0$, T^* prend la valeur de 1, ce qui signifie qu'à la limite de la zone de réaction la température est proche de T_b.

Zone d'équilibre:

Dans la zone d'équilibre, la solution doit satisfaire la condition limite $T^* = 1$ pour $x \to \infty$. La solution triviale est

$$T = 1 \qquad \text{pour} \qquad x > 0\,. \tag{5.39)*}$$

Solution pour y:

On peut ainsi déterminer les conditions limites pour (5.36)*. Les solutions obtenues pour les zones extérieures doivent rejoindre celles de la zone de réaction aux points frontières entre les zones. Nous utiliserons pour cela une coordonnée intermédiaire x_z, et une fonction $k(\varepsilon)$, qui ont la propriété $\varepsilon \ll k(\varepsilon) \ll 1$

$$x_z = \frac{x}{k(\varepsilon)} = \frac{\varepsilon}{k(\varepsilon)} \zeta . \qquad (5.40)^*$$

Les solutions des zones de préchauffage et de réaction se rejoignent pour $\zeta \to -\infty$ dans la zone de réaction et $x \to 0$ dans la zone de préchauffage quand $\varepsilon \to 0$. Quand $\zeta \to -\infty$ dans la zone de réaction, T^* diminue, y augmente. La relation (5.36)* a pour solution à la limite

$$\zeta \to -\infty : \qquad y \to c_0 + c_1 \zeta . \qquad (5.41)^*$$

L'équation différentielle (5.36)* reste inchangée par une translation. En choisissant un changement de coordonnée

$$\zeta = \zeta' - \frac{c_0}{c_1} , \qquad (5.42)^*$$

la solution (5.41)* devient $y \to c_1 \zeta'$. c_0 peut être choisi nul. La solution dans la zone de préchauffage peut être, à proximité de $x = 0$, développée en série de Taylor

$$T = 1 + x \dots . \qquad (5.43)^*$$

Il est seulement nécessaire que la valeur limite

$$\lim_{\varepsilon \to 0} (T_{\text{extérieur}} - T_{\text{intérieur}}) = \lim_{\varepsilon \to 0} \big(1 + x - (1 - \varepsilon\, c_1\, \zeta)\big) \qquad (5.44)^*$$

soit nulle. On a la condition suivante

$$\lim_{\varepsilon \to 0} \big(1 + k(\varepsilon)\, x_z - (1 - c_1\, k(\varepsilon)\, x_z)\big) = 0 , \qquad (5.45)^*$$

qui donne la relation

$$c_1 = 1 . \qquad (5.46)^*$$

On procède de la même manière pour la frontière entre la zone de réaction et la zone d'équilibre. Les conditions limites pour (5.36)* deviennent

$$\zeta \to -\infty : \qquad y \to -\zeta ,$$

$$\zeta \to +\infty : \qquad y \to 0 .$$

Avec $z = dy/d\zeta$ et $d^2y/d\zeta^2 = (1/2)d/dy(z^2)$ on intègre la relation (5.36)* entre $\zeta = -\infty$, $y = -\infty$ et $\zeta = +\infty$, $y = 0$

$$z^2(\infty) = z^2(-\infty) = \Lambda \int_{-\infty}^{0} (y + by^2)\exp(-y)\,dy\,, \qquad (5.47)^*$$

qui donne comme solution pour Λ

$$2\Lambda(1+2b) = 1\,, \qquad (5.48)^*$$

avec $z(0) = 0$ et $z(-\infty) = -1$. En remplaçant (5.31), (5.33) et (5.35)* dans (5.48)*, on obtient une forme qui est identique à la relation (5.22). La solution numérique de la variation de y en fonction de la coordonnée ζ est présentée pour différentes valeurs de b sur la figure 13.

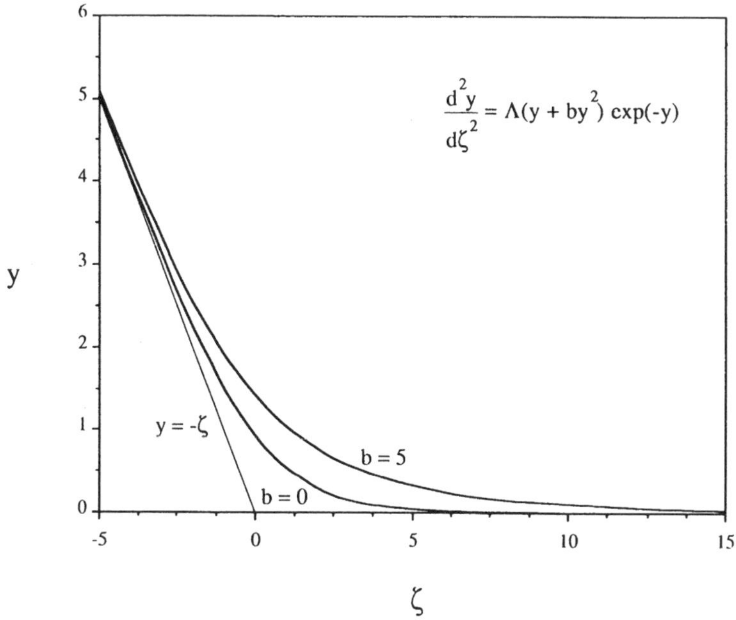

Figure 13:

Variation de y en fonction de la coordonnée étendue ζ.

6. Structure Asymptotique D'une Flamme Méthane-air Stœchiométrique

Reprenons le schéma à quatre réactions globales obtenues au chapitre précédent

$$\begin{array}{ll} \text{I} & CH_4 + 2H + H_2O = CO + 4H_2 \\ \text{II} & CO + H_2O = CO_2 + H_2 \\ \text{III} & 2H + M = H_2 + M \\ \text{IV} & O_2 + 3H_2 = 2H + 2H_2O \end{array} \tag{6.1}$$

Celles-ci ont été obtenues à partir des hypothèses de stationnarité pour les radicaux CH_3, CH_2O, CHO, CH_3O, OH, O et HO_2. Nous ne considérerons ici que les vitesses des réactions élémentaires les plus importantes

$$\begin{array}{ll} w_I = w_{11} \,, & w_{II} = w_{10} \,, \\ w_{III} = w_5 \,, & w_{IV} = w_1 \,. \end{array} \tag{6.2}$$

Les réactions élémentaires correspondantes étant

$$\begin{array}{ll} 11 & CH_4 + H \rightarrow CH_3 + H_2 \\ 10 & CO + OH \rightleftharpoons CO_2 + H \\ 5 & H + O_2 + M \rightarrow HO_2 + M \\ 1 & H + O_2 \rightleftharpoons OH + O \end{array}$$

En supposant l'équilibre partiel des réactions

$$\begin{array}{ll} 2 & O + H_2 \rightleftharpoons OH + H \\ 3 & OH + H_2 \rightleftharpoons H_2O + H \end{array}$$

les vitesses de réaction s'écrivent finalement

$$\begin{aligned} w_I &= k_{11}\, C_{CH_4}\, C_H \\ w_{II} &= \frac{k_{10}}{K_3}\frac{C_H}{C_{H_2}}\left(C_{CO}\, C_{H_2O} - \frac{1}{K_{II}}\, C_{CO_2}\, C_{H_2}\right) \\ w_{III} &= k_5\, C_H\, C_{O_2}\, C_M \\ w_{IV} &= k_1\, \frac{C_H}{C_{H_2}^3}\left(C_{O_2}\, C_{H_2}^3 - \frac{1}{K_{IV}}\, C_H^2\, C_{H_2O}^2\right) \end{aligned} \tag{6.3}$$

où seules les concentrations des 7 espèces intervenant dans (6.1) apparaissent. Les constantes d'équilibre sont approximées par les relations

$$\begin{aligned} K_3 &= 0.216\, \exp(7658/T) \\ K_{II} &= 0.035\, \exp(3652/T) \\ K_{IV} &= 1.48\, \exp(6133/T) \end{aligned} \tag{6.4}$$

Ce schéma peut être réduit à trois réactions globales, en supposant la stationnarité du radical H

$$
\begin{array}{lrcl}
\text{I}' & CH_4 + O_2 & = & CO + H_2 + H_2O \\
\text{II}' & CO + H_2O & = & CO_2 + H_2 \\
\text{III}' & O_2 + 2H_2 & = & 2H_2O
\end{array}
$$

Les vitesses w_I, w_{II} et w_{III} restant valables. La concentration de H est déterminée à partir de la condition $w_I + w_{III} = w_{IV}$

$$
\begin{aligned}
C_H &= C_{H_{eq}} \left(1 - \frac{k_5\, C_M}{k_1} - \frac{k_{11}\, C_{CH_4}}{k_1\, C_{O_2}}\right)^{1/2} , \\
C_{H_{eq}} &= K_{IV}^{1/2} \frac{C_{O_2}^{1/2}\, C_{H_2}^{3/2}}{C_{H_2O}} .
\end{aligned}
\tag{6.5}
$$

Le terme $k_5 C_M / k_1$ peut être négligé devant les autres termes dans (6.5). Nous ferons une hypothèse supplémentaire en supposant que la réaction II est à l'équilibre partielle. Si la température et les concentrations en H_2O et CO_2 peuvent être supposées connues, on a la relation suivante entre les concentrations de CO et H_2

$$
C_{CO} = \alpha'\, C_{H_2} \tag{6.6}
$$

avec

$$
\alpha' = \frac{C_{CO_2}}{C_{H_2O}\, K_{II}(T)} .
$$

En introduisant les opérateurs L et L_i dans les équations (5.7) on obtient

$$
\begin{aligned}
L(T) &\equiv (\rho v)_u \frac{\mathrm{d}T}{\mathrm{d}x} - \frac{\mathrm{d}}{\mathrm{d}x}\left(\frac{\lambda}{c_p}\frac{\mathrm{d}T}{\mathrm{d}x}\right) = \frac{1}{c_p} \sum_{k=I'}^{III'} (-\Delta H_k)\, w_k \\
L_i(Y_i) &\equiv (\rho v)_u \frac{\mathrm{d}Y_i}{\mathrm{d}x} - \frac{1}{\mathrm{Le}_i}\frac{\mathrm{d}}{\mathrm{d}x}\left(\frac{\lambda}{c_p}\frac{\mathrm{d}Y_i}{\mathrm{d}x}\right) = M_i \sum_{k=I'}^{III'} \nu_{ik}\, w_k
\end{aligned}
\tag{6.7}
$$

Le terme $\sum_{i=1}^{n} c_{pi} j_i \mathrm{d}T/\mathrm{d}x$ a été négligé, et c_p est prise constante. Le flux de diffusion j_i s'écrit de la façon suivante

$$
j_i = \rho \mathcal{D}_i \frac{\mathrm{d}Y_i}{\mathrm{d}x} = \frac{1}{\mathrm{Le}_i} \frac{\lambda}{c_p} \frac{\mathrm{d}Y_i}{\mathrm{d}x}
$$

$\mathcal{D}_i$ est le coefficient de diffusion binaire de l'espèce i dans l'azote. Le nombre de Lewis est supposé constant dans tous les cas. Posons

$$
\Gamma_i = \frac{Y_i}{M_i} , \tag{6.8}
$$

on a le système suivant à partir de $(6.7)_2$

$$
\begin{aligned}
L_F(\Gamma_F) &= -w_{I'} \\
L_{O_2}(\Gamma_{O_2}) &= -w_{I'} - w_{III'} \\
L_{H_2}(\Gamma_{H_2}) &= w_{I'} + w_{II'} - 2\,w_{III'} \\
L_{CO}(\Gamma_{CO}) &= w_{I'} - w_{II'} \\
L_{H_2O}(\Gamma_{H_2O}) &= w_{I'} - w_{II'} + 2\,w_{III'} \\
L_{CO_2}(\Gamma_{CO_2}) &= w_{II'}
\end{aligned}
\tag{6.9}
$$

avec $F = CH_4$. $w_{II'}$ peut être éliminée en combinant l'équation pour CO avec les équations de H_2, H_2O, CO_2

$$
\begin{aligned}
L_{H_2}(\Gamma_{H_2}) + L_{CO}(\Gamma_{CO}) &= 2\,w_{I'} - 2\,w_{III'} \\
L_{H_2O}(\Gamma_{H_2O}) - L_{CO}(\Gamma_{CO}) &= 2\,w_{III'} \\
L_{CO_2}(\Gamma_{CO_2}) + L_{CO}(\Gamma_{CO}) &= w_{I'}
\end{aligned}
\tag{6.10}
$$

Il est commode d'utiliser les grandeurs adimensionnelles suivantes

$$
\begin{aligned}
X_i &= \frac{Y_i\,M_F}{Y_{Fu}\,M_i}, & x_i &= \frac{X_i}{\mathrm{Le}_i} \\
T^* &= \frac{T - T_u}{T_b - T_u}, & x^* &= (\rho v)_u \int_0^x \frac{c_p}{\lambda}\,\mathrm{d}x \\
w_k^* &= \frac{\lambda\,M_F\,w_k}{c_p\,Y_{Fu}\,(\rho v)_u^2}, & Q_k^* &= \frac{(-\Delta H_k)\,Y_{Fu}}{c_p\,(T_b - T_u)\,M_F}
\end{aligned}
\tag{6.11)*}
$$

et

$$
\alpha = \frac{X_{CO_2}\,\mathrm{Le}_{H_2}}{X_{H_2O}\,\mathrm{Le}_{CO}\,K_{II}(T)} = \alpha'\,\frac{\mathrm{Le}_{H_2}}{\mathrm{Le}_{CO}}\,. \tag{6.12)*}
$$

Dans la zone de réaction, le terme de convection de l'opérateur L_i peut être négligé devant le terme de diffusion. En supposant de plus α constant dans la zone de réaction, on obtient

$$
\begin{aligned}
-\frac{\mathrm{d}^2}{\mathrm{d}x^2}(x_F) &= -w_{I'} \\
-\frac{\mathrm{d}^2}{\mathrm{d}x^2}(x_{O_2}) &= -w_{I'} - w_{III'} \\
-\frac{\mathrm{d}^2}{\mathrm{d}x^2}(x_{H_2}) &= \frac{2}{1+\alpha}(w_{I'} - w_{III'}) \\
-\frac{\mathrm{d}^2}{\mathrm{d}x^2}(x_{CO}) &= \frac{2\alpha}{1+\alpha}(w_{I'} - w_{III'}) \\
-\frac{\mathrm{d}^2}{\mathrm{d}x^2}(x_{H_2O}) &= \frac{2\alpha}{1+\alpha}w_{I'} + \frac{2}{1+\alpha}w_{III'} \\
-\frac{\mathrm{d}^2}{\mathrm{d}x^2}(x_{CO_2}) &= \frac{1-\alpha}{1+\alpha}w_{I'} + \frac{2\alpha}{1+\alpha}w_{III'}
\end{aligned}
\qquad (6.13)^*
$$

Ce système correspond au mécanisme à deux réactions globales

$$
\begin{aligned}
&\mathrm{I''} \qquad CH_4 + O_2 = \frac{2}{1+\alpha}(H_2 + \alpha CO) + \frac{2\alpha}{1+\alpha}H_2O + \frac{1-\alpha}{1+\alpha}CO_2\,, \\
&\mathrm{III''} \qquad O_2 + \frac{2}{1+\alpha}(H_2 + \alpha CO) = \frac{2}{1+\alpha}H_2O + \frac{2\alpha}{1+\alpha}CO_2\,.
\end{aligned}
$$

La combinaison $(H_2 + \alpha CO)$ apparait ici comme une espèce intermédiaire, formée par la première réaction, et consommée par la seconde. De plus, à partir de $(6.13)^*$ et $(6.7)_2$ on obtient les relations de couplage

$$
\begin{aligned}
&\frac{\mathrm{d}^2}{\mathrm{d}x^2}\left[(1+\alpha)x_{H_2} + 4x_F - 2x_{O_2}\right] = 0 \\
&\frac{\mathrm{d}^2}{\mathrm{d}x^2}\left[x_{H_2} + 2x_F + x_{H_2O}\right] = 0 \\
&\frac{\mathrm{d}^2}{\mathrm{d}x^2}\left[\alpha x_{H_2} + x_F + x_{CO_2}\right] = 0 \\
&\frac{\mathrm{d}^2}{\mathrm{d}x^2}\left[(q_{H_2} + \alpha q_{CO})x_{H_2} + x_F + T\right] = 0
\end{aligned}
\qquad (6.14)^*
$$

qui représentent une combinaison linéaire des équations de conservation des éléments C, H, O et de l'enthalpie h. Les concentrations x_{O_2}, x_{H_2O}, x_{CO_2} et la température adimensionnelle peuvent être calculées, lorsque les équations pour x_F et x_{H_2} dans (6.13) sont résolues. Dans $(6.14)^*$ les expressions

$$
\begin{aligned}
q_{H_2} &= \frac{1}{2}(-\Delta\,\mathrm{H})_{III'}/(-\Delta\mathrm{H}) = 0.3116 \\
q_{CO} &= \left(\frac{1}{2}(-\Delta\,\mathrm{H})_{III'} + (-\Delta\,\mathrm{H})_{II'}\right)/(-\Delta\,\mathrm{H}) = 0.3479
\end{aligned}
\qquad (6.15)^*
$$

sont les chaleurs de formation des espèces H_2 et CO pour les réactions

$$2\,H_2 + O_2 = 2\,H_2O$$

et

$$CO + \frac{1}{2}\,O_2 = CO_2\,,$$

relativement à la réaction totale

$$CH_4 + 2\,O_2 \rightarrow CO_2 + 2\,H_2O\,.$$

Pour plus de simplicité, nous prendrons par la suite $q_{H_2} = q_{CO} = q = 0.33$. Pour l'étude de la structure de la flamme, la zone de flamme est séparée en trois zones distinctes: la zone de préchauffage, la zone de réaction, dans laquelle le combustible est consommé, et H_2 et CO sont formés et la zone d'oxydation du H_2 et du CO. La figure 14 représente schématiquement ces zones.

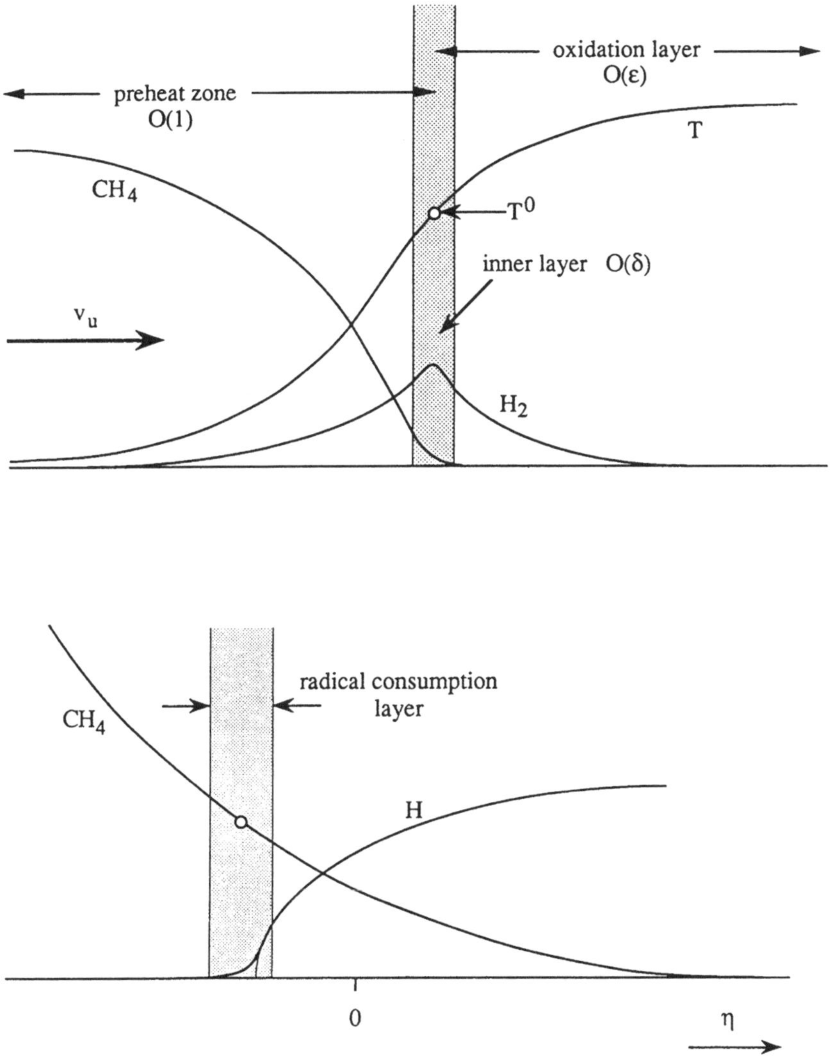

Figure 14:

Structure asymptotique d'une flamme prémélangée.

Dans la zone de préchauffage, ainsi que nous l'avons vu dans le chapitre précédent, le terme de réaction peut être négligé. La relation (6.7) pour CH_4 s'écrit simplement

$$\frac{dX_F}{dx} = \frac{1}{Le_F}\frac{d^2X_F}{dx^2}. \qquad (6.16)^*$$

Les conditions limites sont

$$x \to 0: \qquad X_F = 0,$$

$$x \to -\infty: \qquad X_F = 1.$$

L'intégration de la relation (6.16)* donne

$$X_F = 1 - \exp(Le_F\, x). \qquad (6.17)^*$$

Dans la zone de réaction à laquelle nous nous interessons maintenant, la température et les concentrations de toutes les espèces, exepté CH_4 et H, sont supposées constantes. Pour déterminer la concentration de H, on utilise la relation (6.5), dans laquelle le deuxième terme entre parenthèse peut être négligé pour des pressions pas trop élevées. La vitesse de réaction $w_{I'}$ s'écrit

$$w_{I'} = Da_{I'}\, x_F \left(1 - \frac{x_F}{\delta}\right)^{1/2}. \qquad (6.18)^*$$

Le nombre de Damk"ohler de la première réaction est défini par

$$Da_{I'} = \frac{\rho_0^2}{\rho_u^2\, v_u^2}\frac{Y_{Fu}}{M_F}\frac{\lambda_0}{c_{p0}}\frac{\left(K_{IV}\, X_{O_2}\, X_{H_2}^3\right)_0^{1/2}}{(X_{H_2O})_0}\, Le_F\, k_{11}(T_0). \qquad (6.19)^*$$

L'indice 0 indique que les grandeurs sont prises à $x = 0$. δ est un paramètre petit, défini par

$$\delta = \frac{k_1(T_0)\, X_{O_2,0}}{k_{11}(T_0)\, Le_F}. \qquad (6.20)^*$$

Il fait intervenir le rapport des constantes de vitesse des réactions 1 et 11, et par là-même le rapport entre la production du radical H par la réaction 1, et sa consommation, par la réaction 11. Pour une température typique T_0 (environ 1600 K), k_1 est très petit devant k_{11}, δ est donc petit. D'après la relation (6.18)* nous pouvons voir que la valeur finale de H, et donc de $w_{I'}$, ne peut exister que si $x_{CH_4} \leq \delta$. Ceci n'est vrai que dans la fine zone de réaction. Dans celle-ci, nous définissons la nouvelle coordonnée étendue

$$\zeta = \frac{x}{\delta} \qquad (6.21)^*$$

et

$$y = \frac{x_F}{\delta}.$$

A partir de (6.13)* et (6.18)* on a

$$\frac{\mathrm{d}^2 y}{\mathrm{d}\zeta^2} = (\delta^2 \mathrm{Da}_{I'})\, y\, (1-y)^{1/2} . \tag{6.22)*$$

Nous retrouvons de la même manière qu'au chapitre précédent que le nombre de Damk"ohler doit être de l'ordre de grandeur de $O(\delta^{-2})$, les autres termes étant de l'ordre de 1.

Les conditions limites sont données aux frontières de la zone de réaction avec les zones de préchauffage et d'oxydation

$$\zeta \to +\infty : \qquad y = 0\,, \qquad \frac{\mathrm{d}y}{\mathrm{d}\zeta} = 0\,,$$

$$\zeta = -1 : \qquad y = 1\,, \qquad \frac{\mathrm{d}y}{\mathrm{d}\zeta} = -1\,.$$

Pour cette dernière condition, il suffit de se rapporter à la relation (6.17)*. L'équation (6.22)* peut être intégrée

$$\delta^2\, \mathrm{Da}_{I'} = \frac{15}{8} . \tag{6.23)*$$

La vitesse de combustion est ainsi obtenue par l'intermédiaire de $\mathrm{Da}_{I'}$, si on connait à $x = 0$ les concentrations de O_2, H_2 et H_2O. Il faut pour cela connaître la structure de la zone d'oxydation de H_2 et CO. On a dans cette zone $x_F = 0$ et $w_I = 0$. La concentration de H_2 est la nouvelle variable indépendante

$$x_{H_2} = \frac{\varepsilon\, z}{(1+\alpha)\, q} , \tag{6.24)*$$

où ε est encore un paramètre petit et z la variable indépendante dans la zone d'oxydation. L'intégration des relations de couplage $(6.14)_1^*$ et $(6.14)_4^*$ donnent avec les conditions limites $x_{O_2} = x_{H_2} = 0$ et $T^* = 1$ pour $x \to +\infty$

$$\begin{aligned} x_{O_2} &= \frac{\varepsilon\, z}{2q} \\ T &= 1 - \varepsilon\, z \end{aligned} \tag{6.25)*$$

et la vitesse de réaction $w_{III'}$ s'écrit

$$w_{III'} = 2\, q\, \mathrm{Da}_{III'}\, \varepsilon^3\, z^3 \tag{6.26)*$$

avec

$$\mathrm{Da}_{III'} = \frac{\rho_0^2}{\rho_u^2\, v_u^2} \frac{Y_{Fu}}{M_F} \frac{\lambda_0}{c_{p0}} \left(\frac{K_{IV}\, \mathrm{Le}_{O_2}^3\, \mathrm{Le}_{H_2}^3}{2^5\, (1+\alpha)^3} \right)_0^{1/2} \frac{k_5(T_0)\, C_M}{q^4\, X_{H_2O}} . \tag{6.27)*$$

On peut maintenant connaitre l'évolution de ζ dans cette zone d'oxydation. A partir de $(6.13)_3^*$ ($w_{I'} = 0$ dans cette zone d'oxydation), et avec

$$\eta = \frac{2\,q\,x}{\varepsilon} \tag{6.28)*$$

on obtient

$$\frac{\mathrm{d}^2 z}{\mathrm{d}\eta^2} = \left(\varepsilon^4\,\mathrm{Da}_{III}\right) z^3 . \qquad (6.29)^*$$

Ceci donne comme définition de ε

$$\varepsilon = \mathrm{Da}_{III}^{-1/4} . \qquad (6.30)^*$$

A partir de $(6.14)_2^*$ que l'on intègre, on obtient

$$-\frac{1+\alpha}{2}\left[\frac{\mathrm{d}x_{H_2}}{\mathrm{d}x}\right]_{0_-}^{0_+} = \left[\frac{\mathrm{d}x_F}{\mathrm{d}x}\right]_{0_-}^{0_+} = 1 . \qquad (6.31)^*$$

La deuxième égalité donne la pente du profil de x_F. Dans la zone de réaction, la pente du profil de H_2 est en première approximation nulle, H_2 étant très petit. Avec $x_{H_2} = 0$ pour $x \to \infty$, on obtient pour $(6.29)^*$ les conditions limites

$$\eta = 0: \qquad \frac{\mathrm{d}z}{\mathrm{d}\eta} = -1 ,$$

$$\eta \to \infty: \qquad z = 0 .$$

La solution de l'équation $(6.29)^*$ est

$$z = \frac{2^{1/2}}{2^{1/4} + \eta} . \qquad (6.32)$$

On a ainsi pour $\eta = 0$

$$z_0 = 2^{1/4} . \qquad (6.33)$$

La figure 15 représente la fonction $z(\eta)$. En écrivant le nombre de Damk"ohler Da_{III} de la zone d'oxydation de CO et H_2 pour $\eta = 0$, c'est à dire aussi $x = 0$, on peut trouver une relation simple en combinant Da_I et et $(\mathrm{Da}_{III})_0$: A partir de $(6.23)^*$, on remplace Da_I par $(6.19)^*$ et δ par $(6.20)^*$. Les expressions de x_{O_2} et x_{H_2} sont connues par $(6.24)^*$ et $(6.25)^*$, et écrites pour $x = 0$. La forme obtenue est

$$\frac{k_1^2(T_0)}{k_{11}(T_0)\,k_5(T_0)\,/\,R\,T_0} = 1.5\,p\,\frac{\mathrm{Le}_F}{\mathrm{Le}_{O_2}} . \qquad (6.34)$$

Ici, C_M a été calculé pour une flamme de méthane stœchiométrique en tenant compte des coefficients d'éfficacité des différentes espèces présentes

$$C_M \simeq \frac{1.66\,p}{R\,T} .$$

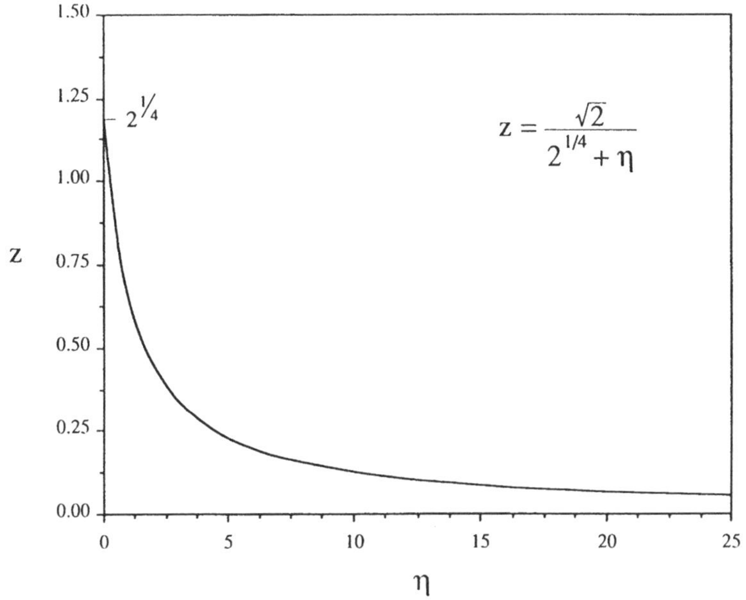

Figure 15:
Evolution de Z en fonction de la coordonnée η.

La relation (6.34) montre que la température de la zone de réaction pour une flamme stœchiométrique de méthane est fonction de la pression, mais indépendante d'autres facteurs extérieurs. En particulier, elle est indépendante de la température des gaz brûlés (à $\eta \to \infty$). Il nous reste encore à déterminer la vitesse de combustion d'une flamme stœchiométrique de méthane: à partir de (6.30)* et avec (6.27)* et (6.34) on obtient l'expression suivante

$$s_F^2 = \frac{8}{15} \frac{k_1^2(T_0)}{k_{11}(T_0)} \frac{\mathrm{Le}_F^{-1} Y_{Fu} \lambda_0}{q^4 X_{H_2O} M_F c_{p0}} \left(\frac{\mathrm{Le}_{O_2}^5 \mathrm{Le}_{H_2}^3 K_{IV}(T_0)}{2^5 (1+\alpha_0)^3} \right)^{1/2} \frac{T_u^2}{T_0^2} \frac{(T_b - T_0)^4}{(T_b - T_u)^4}, \tag{6.35}$$

où

$$\frac{T_b - T_0}{T_b - T_u} = \varepsilon z_0 .$$

s_F peut ainsi être calculé en fonction des paramètres connus (figure 16).

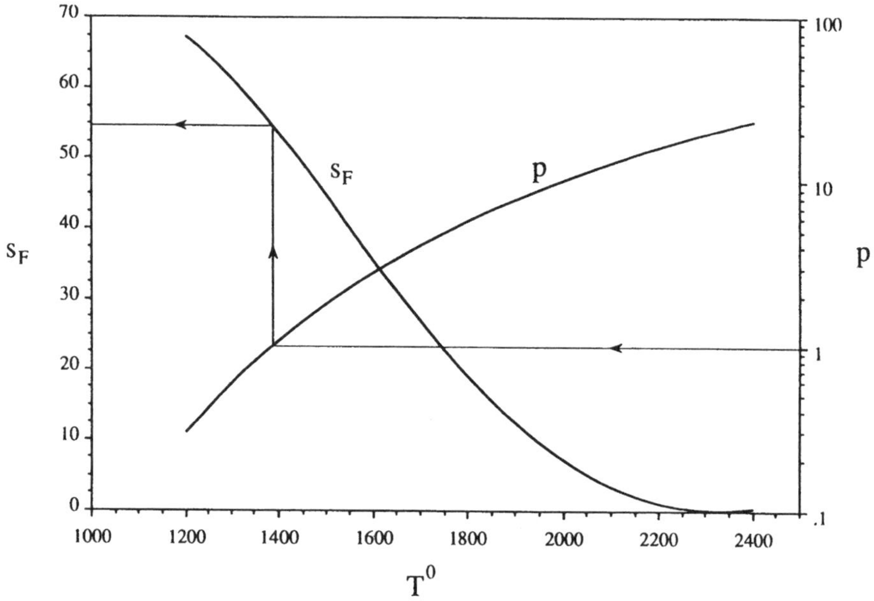

Figure 16:
Variation de la vitesse de flamme et de la pression en fonction de la température de la zone de combustion.

La valeur de α_0 peut-être déterminée comme une fonction de T_0 (6.12)*, avec en première approximation $X_{H_2O} = 2$ et $X_{CO_2} = 1$ dans la zone des gaz brûlés. On remarque, à partir de l'expression (6.35) que la vitesse de flamme diminue jusqu'à devenir nulle pour une valeur de T_b s'approchant de T_0. Cela signifie par exemple, qu'une combustion ne peut avoir lieu que si les pertes de chaleur (faisant diminuer T_b) restent suffisamment faibles pour que $T_b > T_0$.

La relation (6.35) contient encore trop de simplifications, en particulier d'un point de vue cinétique, pour que les valeurs ainsi calculées de la vitesse de flamme soient satisfaisantes. Elle permet cependant de retenir l'influence des paramètres thermodynamiques et cinétiques les plus importants: en particulier, l'influence de la pression ou du rapport des constantes de vitesse k_1^2/k_{11} sur la vitesse de flamme.

SIMULATION OF IGNITION PROCESSES

J. Warnatz
Institut für Technische Verbrennung, Universität Stuttgart

ABSTRACT

A detailed description of the present state of simulation of ignition processes is given, using different complex levels of computation. For this purpose, the instationary conservation equations for total mass, species mass, energy, and momentum are solved, including realistic multi-species models for thermodynamics and transport and detailed chemistry in terms of sets of elementary reactions. Furthermore, suitable numerical methods are given.

The description of ignition processes on different levels includes (a) homogeneous ignition processes and induction times (with applications to engine knock), (b) 1D induced ignition (with application to laser-induced and thermal ignition), (c) 1D auto-ignition and P-T explosion limits, and (d) 2D induced ignition. In all cases, the results are compared to experimental results as far as possible.

1. INTRODUCTION

The simulation of combustion processes needs extreme computational efforts and has therefore to be restricted to relatively simple cases, due to a very complex interaction of chemical reaction, transport, and convection described by instationary or stationary partial differential equation systems with temperature, species concentrations, and flow velocity as dependent variables. Furthermore, already in the simple case of combustion of a small hydrocarbon in one-dimensional geometry, about 50 species involved in 500 reactions have to be considered, resulting in large non-linear stiff partial differential equation systems.

Nevertheless, even a rather simple treatment can give much insight into ignition processes, as shall be demonstrated in the following.

After an introducing section on physical chemistry of combustion processes, simulation of ignition processes is considered for three different cases:

1. *Ignition Delay Time in Auto-Ignition Processes* (e.g. in shock tubes or Otto engines): The simulation of homogeneous (i.e. 0D problems leads to very large non-linear, stiff ordinary differential equation systems which are solved by implicit methods. Large reaction systems (up to ~1500 reactions of ~150 species) can be considered without difficulty.

2. *P-T-Ignition Limits of Auto-Ignition Processes* (e.g. in simple spherical ignition bombs): The simulation of these at least 1D instationary problems leads after spatial discretization to differential/algebraic equation systems which are then solved using both backward differencing formula (BDF) and extrapolation codes. Much smaller mechanisms than for the 0D case have to be used with the computers presently available.

3. *Induced Ignition and Instationary Flame Propagation* (e.g. in spherical or cylindrical ignition bombs): In this case, spatial discretization using finite differences again leads to systems of differential/algebraic equations which are solved by implicit methods mentioned above. The main problem (both for 1D and 2D problems) is the proper construction of adaptive, moving grid point systems in order to treat the flame front as well as the acoustic wave propagation occurring in ignition processes.

2. PHYSICAL CHEMISTRY OF COMBUSTION PROCESSES

2.1 Thermodynamic Data

Thermodynamic data are needed to calculate specific heat capacities and enthalpies in the energy balance equation and for the computation of reverse reaction rate coefficients in the reaction mechanism (see below). For this purpose, polynomial fits of species molar heat capacities $C_{p,i}$ are used, consisting of seven coefficients for high (T > 1000 K) and low (T < 1000 K) temperature. These are in each case five coefficients for the molar heat capacities $C_{p,i}$, and two integration constants for enthalpy and entropy; details can be found in the literature [1-3]. Some examples are presented in Table 1.

Table 1: Examples of Polynomial Representation of Thermodynamic Properties

```
O                 J 3/77O   100   000   000   0G    300.000   5000.000        R1
 0.25342961E+01-0.12478170E-04-0.12562724E-07 0.69029862E-11-0.63797095E-15    2
 0.29231108E+05 0.49628591E+01 0.30309401E+01-0.22525853E-02 0.39824540E-05    3
-0.32604921E-08 0.10152035E-11 0.29136526E+05 0.26099342E+01                   4

O2                J 3/77O   20    00    00    0G    300.000   5000.000         1
 0.36122139E+01 0.74853166E-03-0.19820647E-06 0.33749008E-10-0.23907374E-14    2
-0.11978151E+04 0.36703307E+01 0.37837135E+01-0.30233634E-02 0.99492751E-05    3
-0.98189101E-08 0.33031825E-11-0.10638107E+04 0.36416345E+01                   4

O3                J 6/61O   30    00    00    0G    300.000   5000.000        R1
 0.54665239E+01 0.17326031E-02-0.72204889E-06 0.13721660E-09-0.96233828E-14    2
 0.15214096E+05-0.34712616E+01 0.24660617E+01 0.91703209E-02-0.49698480E-05    3
-0.20634230E-08 0.20015595E-11 0.16059556E+05 0.12172130E+02                   4

N                 J 6/77O   1N    10    00    0G    300.000   5000.000        R1
 0.28897814E+01 0.10005879E-02-0.22048807E-06 0.20191288E-10-0.39409831E-15    2
 0.38857042E+04 0.55566427E+01 0.38737300E+01-0.13393772E-02 0.16348351E-05    3
-0.52133639E-09 0.41826974E-13 0.35802348E+04 0.34202406E+00                   4

N2                J 3/77N   2N    00    00    0G    300.000   5000.000        R1
 0.28532899E+01 0.16022128E-02-0.62936893E-06 0.11441022E-09-0.78057465E-14    2
-0.89008093E+03 0.63964897E+01 0.37044177E+01-0.14218753E-02 0.28670392E-05    3
-0.12028885E-08-0.13954677E-13-0.10640795E+04 0.22336285E+01                   4
```

Mixture specific enthalpies h and heat capacities c_p are then determined from arithmetic means using the mass fractions w_i as weight.

2.2 Transport Models

Diffusive processes take care of broadening of steep profiles, where the temporal change of the quantity considered (e.g. temperature T, mass fractions w_i, velocity v, ...) are proportional to the curvature of the actual profile (2nd law of *Fourier* for T, 2nd law of *Fick* for w_i, z is the space coordinate):

$$\partial X/\partial t = D \cdot \partial^2 X/\partial z^2$$

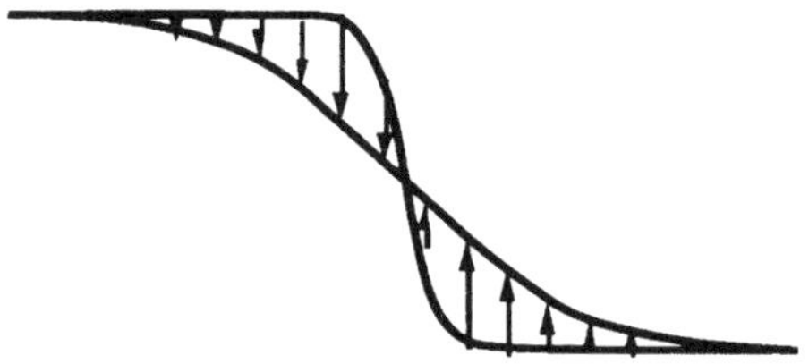

The proportionality coefficient D (i.e. diffusion coefficient D for w_i, heat conductivity λ for T, viscosity η for v, ...) can be determined from intermolecular potential properties like e.g. the Lennard-Jones parameters σ (molecular diameter) and ε (potential well depth).

An really exhausting treatment of this topic is given in the book of *Hirschfelder, Curtiss, and Bird* [4]; a computer package for the determination both of pure species and mixture transport properties has recently been described in the literature [5,6].

2.3 Temperature Dependence of Reaction Rate Coefficients

The temperature dependence of rate coefficients cannot be described by a simple Arrhenius form (resulting from the fact that an energy barrier of height E_a has to be passed)

$$k = A \cdot \exp(-E_a/RT)$$

as it was usually used in the literature up to the seventies (k = rate coefficient, A = pre-exponential factor, E_a = activation energy, R = gas constant, T = temperature). In any case, curved Arrhenius plots described in the simplest case by

$$k = A \cdot T^b \cdot \exp(-E_a/RT)$$

have to be considered, where the additional curvature caused by the temperature coefficient b of the pre-exponential (the measurement of which is very difficult due to the overwhelming influence of the exponential temperature dependence) makes the extrapolation from experimentally accessible to high temperatures rather difficult; see [7] for details.

2.4 Pressure Dependence of Reaction Rate Coefficients

The quantitative description of the behaviour of pressure-dependent reactions is a challenging problem because of its very close connection with the temperature dependence and distribution of product channels; see e.g. [8,9].

The simplest description is the *Lindemann* theory, assuming one-step activation and deactivation of the molecule A considered by a collision partner M to form then an activated molecule A^* which then can undergo a unimolecular decomposition:

$$A + M \rightarrow A^* + M \qquad (k_a)$$
$$A^* + M \rightarrow A + M \qquad (k_{-a})$$
$$A^* \rightarrow P \qquad (k_u)$$

Assumption of quasi-steady-state behaviour of A^* leads to the rate expression

$$d[P]/dt = k_u/(k_{-a}[M] + k_u) \cdot k_a [A] [M]$$

for the formation of products P. For the high pressure and the low pressure range this simplifies to the expressions

HP: $d[P]/dt = k_u/k_{-a} \cdot k_a [A]$ (1st order reaction)

LP: $d[P]/dt = k_a [A] [M]$ (2nd order reaction)

with rate coefficients $k_\infty = k_u\, k_a/k_{-a}$ (high pressure HP) and $k_0 = k_a$ (low pressure LP), as it is shown in Fig. 1, where the ratio

$$k_{exp}/k_\infty = k_{-a} [M] / (k_{-a}[M]+k_u)$$

is plotted as function of the pressure, showing distinctly the transition from first order to second order behaviour. It is plausible that the transition range moves to higher pressures if the temperature is increased and if the molar mass of the collision complex is decreased.

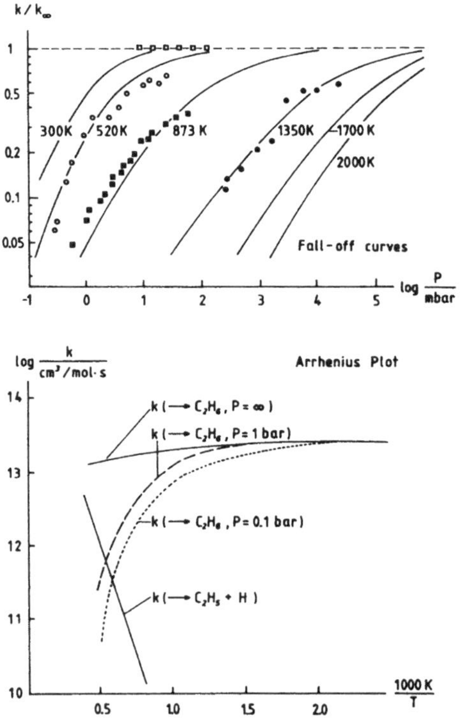

Fig. 1: Fall-off behaviour of rate coefficients for CH_3 combination[10])

More sophisticated treatments take into consideration that activation and deactivation proceed via many collisions, leading to a broadening of the fall-off range. A calculation using "*quantum mechanical Kassel sums*" [10,11]) leads to fall-off curves given in Fig. 1a; Fig. 1b shows the resulting temperature dependence of rate coefficients, if the fall-off curves are combined with the high pressure rate coefficients k_∞.

A more easy treatment is description of the deviation from the Lindemann behaviour by a broadening factor, which has to be determined (usually as a three-parameter expression) for each reaction considered [8,9]).

2.5 Multi-Scale Character of Reaction Systems, Stiffness

As example for the determination of the so-called degree of stiffness the simple reaction sequence

$$S_1 \xrightarrow{k_{12}} S_2 \xrightarrow{k_{23}} S_3$$

shall be considered. The reaction rates r_i for the corresponding concentrations are given by

$$r_1 = d[S_1]/dt = - k_{12}\,[S_1] \quad ; \quad r_2 = d[S_2]/dt = + k_{12}\,[S_1] - k_{23}\,[S_2]$$

The Jacobian matrix J = (dr/d[S]) resulting from this system of ordinary differential equations is then given by

$$J = \begin{pmatrix} -k_{12} & 0 \\ +k_{12} & -k_{23} \end{pmatrix}$$

with the eigenvalues $-k_{23}$ and $-k_{12}$. The degree of stiffness is the ratio of largest and smallest decaying eigenvalue; here it is k_{23}/k_{12}. With a conventional (explicit) solution procedure, the time step has to be reduced by degree of stiffness which easily can reach values like 10^{10}, making solution of the problem impossible.

3. SIMULATION OF THE CHEMISTRY OF AUTO-IGNITION

3.1 Background

Engine knock is a decades old problem connected with the operation of Otto engines, in special since the development of high performance versions (e.g. for airplanes). Very early it became apparent, that knocking is kinetically controlled, and must be closely connected with the chemistry of combustion, as can be easily seen for instance from the strong effect of addition of traces of inhibitors like lead tetraethyl; for reference see [12]). Extension of reaction schemes of the high temperature oxidation of hydrocarbons [13-17]) to lower temperatures allows the explanation of ignition delay up to octane using sophisticated mechanisms for the alkyl radical decomposition of higher hydrocarbons [18-20]), as will be demonstrated in the following.

3.2 Calculation Method

Calculation of ignition delay times (or induction times) is performed by solving the conservation equations of the reaction system, considering the respective experimental conditions of pressure, volume, and temperature which, in some cases (e.g. end-gas of Otto engines), are time-dependent. Treating homogeneous mixtures, these equations represent a system of ordinary differential equations with the time t as independent variable and with the species concentrations c_i as dependent variables. The equation system may be written as

$$\dot{c}_i = \omega_i + c_i \cdot \left\{ \frac{\dot{p}}{p} - \frac{RT \cdot \Sigma\dot{\omega}_i}{p} - \frac{\dot{T}}{T} \right\} \qquad ; \qquad i = 1,\dots,n$$

if the temperature T and the pressure P are given, or in the alternative form

$$\dot{c}_i = \omega_i - c_i \cdot \frac{\dot{V}}{V} \qquad ; \qquad i = 1,\dots,n$$

if the temperature T and the volume V are given, with ω_i = molar rate of formation of species i, R = gas constant, n_s = number of species (typically ~50); $c_i(t=t_0)=c_{i0}$ are the initial conditions of the problem considered.

For the given system of ordinary differential equations for the concentrations and their derivatives

$$\vec{c} = (c_1,\ldots,c_{n_s})^T \quad ; \quad \vec{c}'(t) = f[t,\vec{c}(t),\vec{p}]^T$$

with the system parameters p_j, j=1,...,m (i.e. for example the rate coefficients, if homogeneous chemistry is considered), differentiation with respect to the parameter vector $\vec{p}$ yields then the linear equation system for the (n_s x m) matrix S of sensitivity coefficients $S_{ij}(t) = \partial c_i(t)/\partial p_j$

$$S'(t) = J(t) \cdot S(t) + f_p(t)$$

Here J(t) is the Jacobian matrix and $f_p(t)$ are the n_s x m matrix of local parametric derivatives which easily can be evaluated by numerical differentiation:

$$f_{pij}(t) = \left(\frac{\partial f_i}{\partial p_j} \right)_{p_{k \neq j}, c}$$

with initial values which are given by the expressions at the start time of integration

$$S(t = t_0) = \frac{\partial c_0(p)}{\partial p} \quad ; \quad c_0 = c(t = 0)$$

Solution of this initial value problem is obtained by use of the programs *DASSL* [21] or *LIMEX* [22,23], both of them designed to solve stiff differential/algebraic systems. Solution of the additional sensitivity equations is performed simultaneously with the solution of the model equations [24,25].

3.3 High Temperature Reaction Mechanism of Alkane Combustion

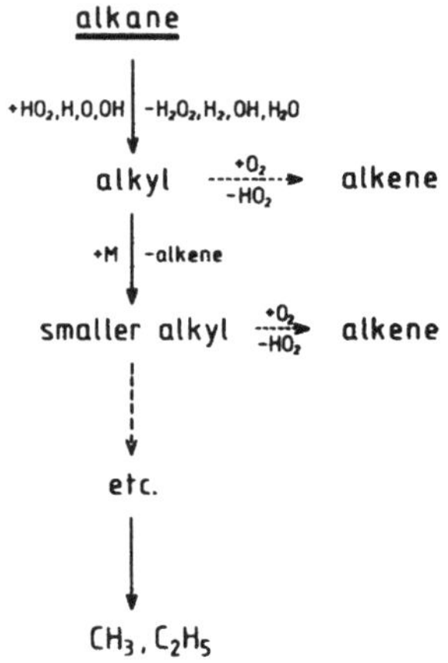

Fig. 2a: Alkyl radical decomposition mechanism

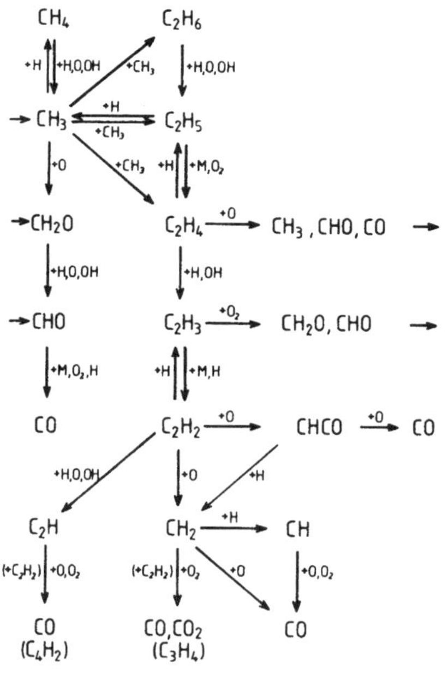

Fig. 2b: Oxidation of C_1/C_2 hydrocarbons

The reaction mechanism used here for the description of the oxidation of molecules up to C_8-species is very similar to that discussed in [13-17]. This reaction mechanism contains a complete oxidation scheme for hydrocarbons up to C_4-species [14]; see Fig.2.

Table 2: Mechanism for H_2/CO Oxidation, P = 1 bar (Oct. 88)

					A(cm,mol,s)	b	E(kJ/mol)	No.
******************************								
**** 1. H2-O2 Mechanism *								
******************************								
---- 1.1 H2-O2 Chain Reactions								
O2	+H	=OH	+O		2.20E+14	0.00	70.3	(1, 2)
H2	+O	=OH	+H		5.06E+04	2.67	26.3	(3, 4)
H2	+OH	=H2O	+H		1.00E+08	1.60	13.8	(5, 6)
OH	+OH	=H2O	+O		1.50E+09	1.14	0.4	(7, 8)
---- 1.2 Recombination Reactions								
H	+H	+M*	=H2	+M*	1.80E+18	-1.00	0.0	(9, 10)
H	+OH	+M*	=H2O	+M*	2.20E+22	-2.00	0.0	(11, 12)
O	+O	+M*	=O2	+M*	2.90E+17	-1.00	0.0	(13, 14)
---- 1.3 HO2 Formation/Consumption								
H	+O2	+M*	=HO2	+M*	2.30E+18	-0.80	0.0	(15, 16)
HO2	+H	=OH	+OH		1.50E+14	0.00	4.2	(17, 18)
HO2	+H	=H2	+O2		2.50E+13	0.00	2.9	(19, 20)
HO2	+H	=H2O	+O		3.00E+13	0.00	7.2	(21, 22)
HO2	+O	=OH	+O2		1.80E+13	0.00	-1.7	(23, 24)
HO2	+OH	=H2O	+O2		6.00E+13	0.00	0.0	(25, 26)
---- 1.4 H2O2 Formation/Consumption								
HO2	+HO2	>H2O2	+O2		2.50E+11	0.00	-5.2	(27)
OH	+OH	+M*	=H2O2	+M*	3.25E+22	-2.00	0.0	(28, 29)
H2O2	+H	=H2	+HO2		1.70E+12	0.00	15.7	(30, 31)
H2O2	+H	=H2O	+OH		1.00E+13	0.00	15.0	(32, 33)
H2O2	+O	=OH	+HO2		2.80E+13	0.00	26.8	(34, 35)
H2O2	+OH	=H2O	+HO2		5.40E+12	0.00	4.2	(36, 37)
******************************								
**** 2. CO-CO2 Mechanism *								
******************************								
---- 2.1 CO-CO2 Reactions								
CO	+OH	=CO2	+H		4.40E+06	1.50	-3.1	(38, 39)
CO	+HO2	=CO2	+OH		1.50E+14	0.00	98.7	(40, 41)
CO	+O	+M*	=CO2	+M*	7.10E+13	0.00	-19.0	(42, 43)
CO	+O2	=CO2	+O		2.50E+12	0.00	200.0	(44, 45)

k = A T**b exp(-E/RT)

[M] = total concentration

[M*] = [H2]+6.5[H2O]+0.4[O2]+0.4[N2]+0.75[CO]+1.5[CO2]+3.0[CH4]

3.4 Low Temperature Reaction Mechanism

Ignition mainly is governed by chain branching processes which are rather unspecific of the fuel considered at high temperature, whereas they are rather fuel specific at low temperature, leading to very complex reaction systems if the different isomeric structures are taken into consideration.

Chain branching at high temperature ($T > 1100$ K at $P = 1$ bar) is predominantly provided by the reaction well known from flame propagation

$$H\cdot + O_2 \rightarrow OH\cdot + O\cdot$$

At medium temperature (1100 K $> T > 900$ K at $P = 1$ bar), additional chain branching is introduced by the reaction sequence given by the steps

$$HO_2\cdot + RH \rightarrow H_2O_2 + R\cdot$$
$$H_2O_2 + M \rightarrow OH\cdot + OH\cdot + M$$

In the low temperature range ($T < 900$ K at $P = 1$ bar), there is so-called "*degenerated chain-branching*", characterized by the fact that the precursor of chain-branching (here: $RO_2\cdot$) is decomposed at higher temperature, leading to an inverse temperature dependence of reaction rate.

$$R\cdot + O_2 \leftrightarrow RO_2\cdot \qquad (O_2 \text{ addition})$$

$$RO_2\cdot + RH \rightarrow RO_2H + R\cdot \qquad (\text{external abstraction})$$
$$RO_2H \rightarrow RO\cdot + OH\cdot \qquad (\text{chain branching})$$

$$RO_2\cdot \rightarrow R'O_2H\cdot \qquad (\text{internal abstraction})$$
$$R'O_2H\cdot \rightarrow R'O + OH\cdot \qquad (\text{chain propagation})$$
$$R'O_2H\cdot \rightarrow R' + HO_2\cdot \qquad (\text{chain propagation})$$

$$R'O_2H\cdot + O_2 \leftrightarrow O_2R'O_2H\cdot \qquad (2^{nd}\ O_2 \text{ addition})$$

3.5 Results at Isothermal Conditions

By solving numerically the conservation equations for a 0D homogeneous isothermal system with chemical reaction, ignition delay times are calculated for several higher hydrocarbons up to octane and compared to experimental shock tube results.

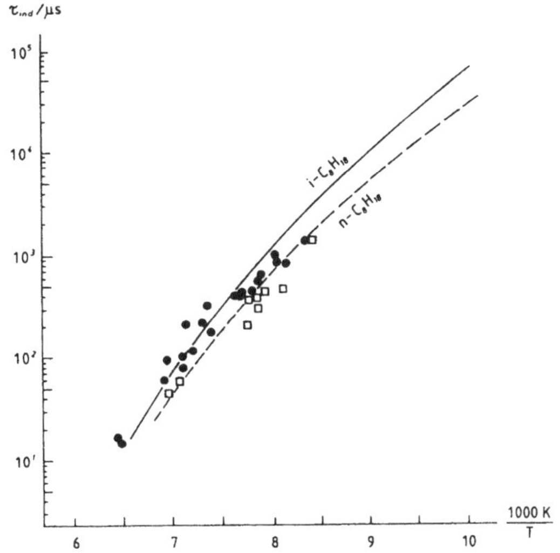

Fig.3: Experimental results (points) and simulation (line) of ignition delay time t_{ind} of i-C_8H_{18} and n-C_8H_{18} in O_2/Ar [27])

The differences in ignition delay result from the pattern of alkyl radical decomposition that changes from one fuel to another. Fig.3 shows a comparison between experimental data and results of corresponding simulations.

Fig.4 gives an example of a sensitivity analysis with respect to OH formation during the ignition delay phase, showing the important rate-limiting role of H_2O_2 chemistry, the OH + alkene reactions, C_8H_{17} decomposition and reaction with O_2 (due to the degenerate branching), and initial attack on the parent hydrocarbon.

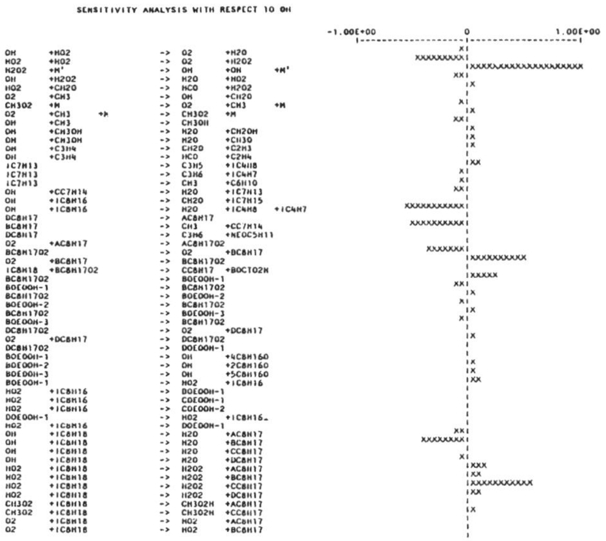

Fig.4: Sensitivity analysis with respect to the concentration of OH for an i-$C_8H_{18}/O_2/Ar$ mixture, T = 800 K

3.6 Results at Changing Temperature and Pressure

The example considered here (Fig.5) is knock in an engine driven with n-C_8H_{18} [28]). Both knocking and non-knocking case can be predicted with a mechanism described above.

4. SIMULATION OF INSTATIONARY 1D-COMBUSTION

4.1 Background

A detailed knowledge of instationary combustion processes is required for the understanding of phenomena like ignition or extinction which are of importance in many practical applications (Otto engines, safety considerations). Time-dependent one-dimensional simulations have to be used to treat such instationary combustion processes. They are performed by solving the corresponding conservation equations (i.e. conservation of mass, energy, momentum and species mass) without restriction to systems with spatially uniform pressure, therefore allowing to study the interesting interaction of pressure waves with the flame front and the influence of the uniform pressure assumption.

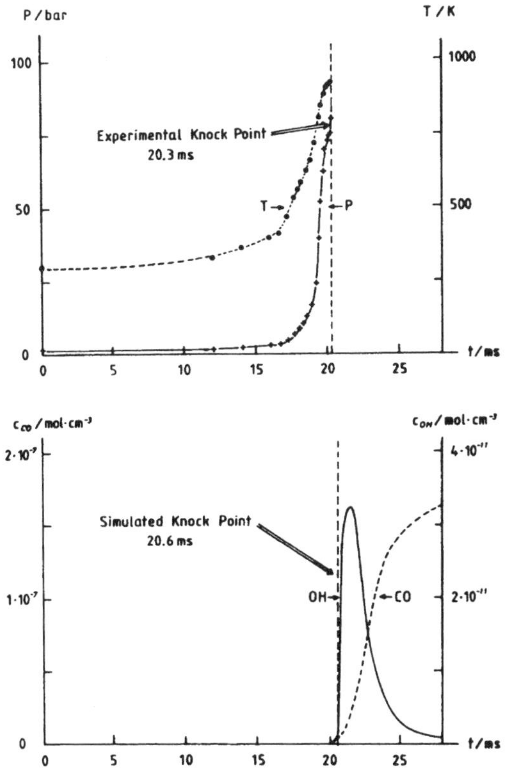

Fig.5: Measured (upper drawing) and calculated (lower drawing) knocking behaviour in an engine fueled with n-octane [28])

Spatial discretization using finite differences and an adaptive grid point system lead to a differential/algebraic equation system which is solved numerically by both extrapolation and backward differencing codes [21-23]. Computer simulations are used to study the thermal ignition of ozone-oxygen as well as hydrogen-oxygen mixtures. As comparisons with corresponding experiments show [29-31], minimum ignition energies in O_3-O_2 mixtures can be simulated well within the experimental error limits.

4.2 Calculation Method

Quantitative mathematical simulation of the ignition process is performed by solving the corresponding system of conservation equations. After transformation into Lagrangian coordinates, the instationary one-dimensional conservation equations may be written in the following form:

$$\frac{\partial r}{\partial \psi} - \frac{1}{\rho r^{\alpha}} = 0$$

$$\frac{\partial \rho}{\partial t} + \rho^2 \frac{\partial}{\partial \psi}(v r^{\alpha}) = 0$$

$$\frac{\partial v}{\partial t} + r^{\alpha}\frac{\partial P}{\partial \psi} - \frac{4}{3} r^{\alpha}\frac{\partial}{\partial \psi}\left(\rho\mu \frac{\partial}{\partial \psi}(v r^{\alpha})\right) + 2\alpha r^{\alpha}\frac{v}{r}\frac{\partial \mu}{\partial \psi} = 0$$

$$\frac{\partial T}{\partial t} - \frac{1}{\rho c_p}\frac{\partial P}{\partial t} - \frac{1}{c_p}\frac{\partial}{\partial \psi}\left(\rho r^{2\alpha}\lambda \frac{\partial T}{\partial \psi}\right) + \frac{r^{\alpha}}{c_p}\sum_{i=1}^{n_s} \rho w_i V_i c_{pi}\frac{\partial T}{\partial \psi}$$
$$+ \frac{1}{\rho c_p}\sum \dot{\omega}_i h_i M_i - \frac{4\rho\mu}{3 c_p}\left(\frac{\partial v r^{\alpha}}{\partial \psi}\right)^2 + \frac{2\alpha\mu}{c_p}\frac{\partial}{\partial \psi}(v r^{\alpha-1}) = q$$

$$\frac{\partial w_i}{\partial t} + \frac{\partial}{\partial \psi}(\rho r^{\alpha} w_i V_i) - \frac{\dot{\omega}_i M_i}{\rho} = 0$$

$$P - \frac{\rho R T}{\overline{M}} = 0$$

with t = time, α = 0 for infinite slab, α = 1 for infinite cylinder, α = 2 for sphere, n_s = number of species, r = radius, P = pressure, T = temperature, ρ = density, v = velocity, w_i = mass fraction of species i, M_i = molar mass of species i, ω_i = mass scale rate of formation of species i, q = source term for deposition of energy, c_{pi} = constant pressure specific heat capacity of species i, c_p = constant pressure specific heat capacity of the mixture, h_i = specific enthalpy of species i, V_i = diffusion velocity of species i, λ = thermal conductivity of the mixture. The independent variables are: t and Ψ; the dependent variables are: ρ, T, P, w_i, v, and r.

The calculations are simplified to a rather great extent, if the pressure in the reaction volume is assumed to be uniform in space. In this case, the momentum equation then is replaced by the equation $\partial P/\partial r = 0$, and the continuity equation (automatically fulfilled after transformation into Lagrangian coordinates) is no longer needed.

The source term in the energy conservation equation is used to simulate ignition by external energy sources, corresponding to the heating by a laser beam in the experiments [29-31]). For spherical and cylindrical geometries at the center of the reaction vessel symmetry conditions are used. Outer boundary conditions for the temperature and the species mass fractions in the simulations of ignition by external energy sources are simplified by assuming zero gradients of temperature and mass fractions; see in the literature [32]) for details.

Dealing with auto-ignition in closed vessels, one has to take into consideration reactions occurring at the surface, like recombination of atoms or destruction of reactive molecules as well as energy transfer to the vessel. The temperature at the outer boundary is assumed to be constant (thermostated in the experiments), and the system is considered at constant volume. Considering the surface reactions

$$\alpha_{lk}A_l \xrightarrow{\gamma_k} \tilde{\alpha}_{1k}A_1 + \tilde{\alpha}_{2k}A_2 + \ldots + \tilde{\alpha}_{n_sk}A_{n_s}$$

occurring at the surface, the outer boundary conditions for species mass conservation are given by the experession (balancing surface reaction and diffusion to the surface)

$$0 = j_i^{\,s} = \rho w_i V_i + \dot{\omega}_i^{\,s} \quad ; \quad \dot{\omega}_i^{\,s} = \sum_{k=1}^{n_s} \gamma_k Z_l M_i (\tilde{\alpha}_{ik} - \delta_{li}\alpha_{lk})$$

with γ_k = surface destruction efficiency, Z_l = surface collision number of species l, α_{ik} = stoichiometric coefficient of species i in reaction k, n_s = number of surface reactions, ω_i^s = mass scale rate of formation of species i per surface unit.

A system of coupled ordinary differential/algebraic equations which can be solved numerically is obtained by spatial discretization using finite differences. For the simulation of ignition by artificial energy sources, adaptive gridding has to be used due to the large ratio of vessel diameter to flame front thickness and to the diameter of the artificial energy source.

A static approach is used which controls the grid point system after each successful step of the time integration by computing a new mesh which is compared with the old one. The grid point density is determined by equipartitioning the integral of a mesh function and inverse interpolation, the mesh function F given by a weighted norm of gradients and curvature of the dependent variables f_m

$$F(\psi) = \sum_{m=1}^{M} a_m \frac{\int_0^{\psi} \left|\frac{\partial f_m}{\partial \psi}\right| d\psi}{\int_0^{\psi_0} \left|\frac{\partial f_m}{\partial \psi}\right| d\psi} + b_m \frac{\int_0^{\psi} \left|\frac{\partial^2 f_m}{\partial \psi^2}\right| d\psi}{\int_0^{\psi_0} \left|\frac{\partial^2 f_m}{\partial \psi^2}\right| d\psi}$$

with a_m, b_m = weighting factors. If necessary, the adaption of the mesh is performed by insertion and deletion of grid points in the old mesh.

This method is chosen in order to minimize the error associated with the interpolation of the solution onto the new mesh. Piecewise monotonic cubic hermite interpolation [33] is then used for the interpolation to avoid overshoots known from spline interpolation.

Simulations without the uniform pressure assumption are complicated by the problem of resolving the shock fronts. In order to avoid here numerical instabilities resulting if shocks are not resolved by the grid point system, an artificial viscosity term (smart numerical diffusion), as it is proposed by *Richtmyer and Morton* [34] is introduced, which spreads the shocks over some grid points.

The system of ordinary differential/algebraic equations is solved using the packages *DASSL* [21] or *LIMEX* [22,23], using implicit integration methodes.

The block-tridiagonal structure allows an efficient numerical evaluation of the Jacobian. As the integration has to be restarted after each adaption of the grid point system, the one-step method *LIMEX* needs less computing time.

Typical cpu-times on a *CRAY-1* computer range from less than 1 minute (simulation without regridding and using the uniform pressure assumption) to more than 3 hours (simulation of ignition and flame propagation without the uniform pressure assumption).

4.3 Results on Induced Ignition and Minimum Ignition Energies

As a comparison between experimental [29-31] and computed results shows (Fig.6), the mathematical model described above is able to calculate the minimum ignition energies in ozone-oxygen mixtures. For hydrogen-oxygen mixtures, only computational results are presented. An example of a simulation of an igniting hydrogen-oxygen mixture (without using the uniform pressure assumption) is shown in Figs. 7 and 8 for typical values of ignition source time t_s, ignition source diameter r_s, and ignition source energy E_s.

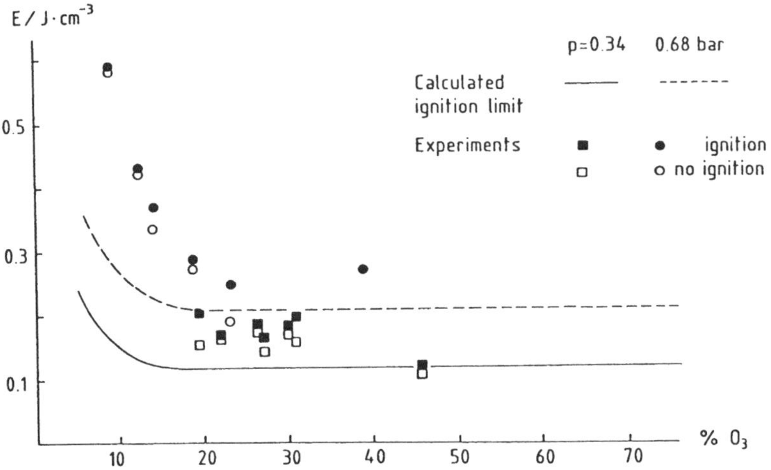

Fig.6: Measured (points) and calculated (lines) minimum thermal ignition energies in O_3-O_2 mixtures at 0.34 and 0.68 bar; unburnt gas temperature T_u = 298 K, ignition source time t_s = 1 ms

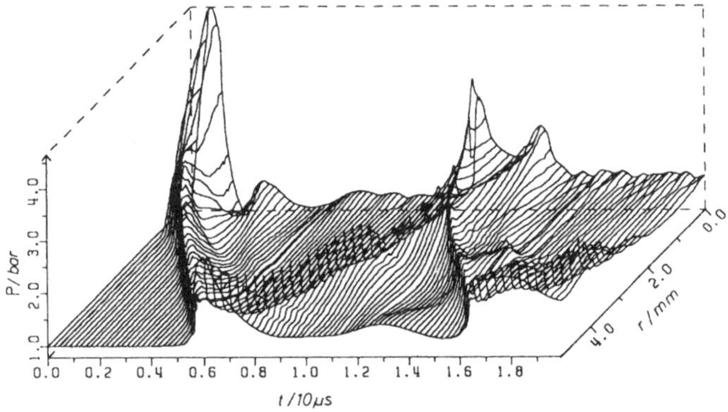

Fig.7: Calculated pressure profiles in an igniting hydrogen-oxygen mixture (stoichiometric); cylindrical geometry, P = 1 bar, t_s = 1 ms, r_s = 1 mm, E_s = 4 J

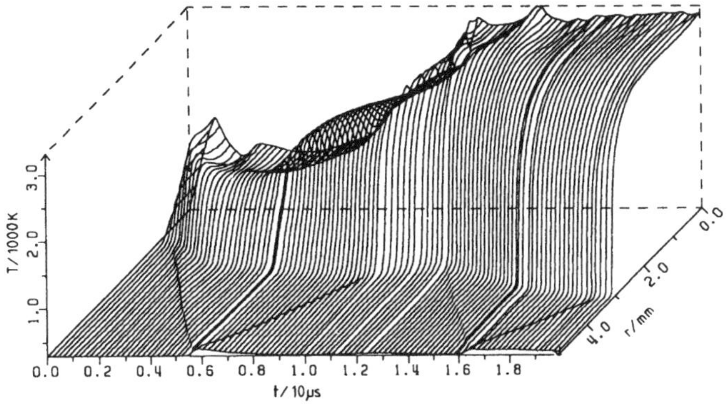

Fig.8: Calculated temperature profiles in an igniting stoichiometric hydrogen-oxygen mixture; cylindrical geometry, P = 1 bar, t_s = 1 ms, r_s = 1 mm, E_s = 4 J

For short ignition times (<10 ms), the ignition process may be characterized as follows: In the heating period there is a quick rise of temperature and pressure in the source volume, the time being too short for the pressure to distribute over the whole volume of the mixture. Then a diverging pressure wave and a converging rarefaction wave are formed.

The rarefaction wave is reflected at the vessel center forming a diverging wave, the shock wave (moving in direction of the outer boundary) is reflected, forming a converging shock. The pressure waves perturb the propagation of the flame front, and the overall movement of the flame may be described as a superposition of the "normal" flame propagation and the oscillatory movements caused by the crossing shocks.

If ignition times are long in comparison to the time scale of the gas-dynamic processes, there is enough time for the pressure to equilibrate in the reaction system during the heating of the source volume, and the pressure is nearly uniform in space, which means that the uniform pressure assumption should be a rather good approximation no differences in the computed minimum ignition energies for an ignition time of 100 ms; for further reference on this topic see [32]).

Fig.9 shows the dependence of minimum ignition energies on the energy source radius for cylindrical geometry in a hydrogen-oxygen-mixture at three different ignition times. The slope of the broken line indicates proportionality between the ignition energy and source volume, i.e. the energy density for ignition (or ignition temperature) is nearly constant. For smaller radii, diffusion and heat conduction cause the energy densities necessary for ignition to increase.

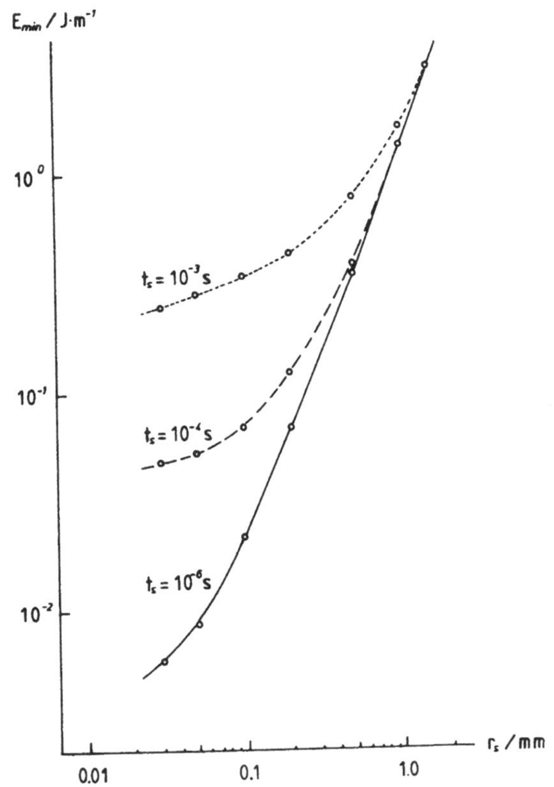

Fig.9: Calculated ignition energies in 2:1 hydrogen-oxygen mixtures for different radii of the external energy source (P = 1 bar)

The variation of ignition energy with mixture composition is shown in Fig. 10. There is almost no dependence on mixture composition for a source radius of 1 mm between the flammability limits. For small ignition radii (here 0.2 mm) the ignition energies become dependent on the mixture composition because rich mixtures (with higher thermal conductivity) need higher ignition energies than lean mixtures.

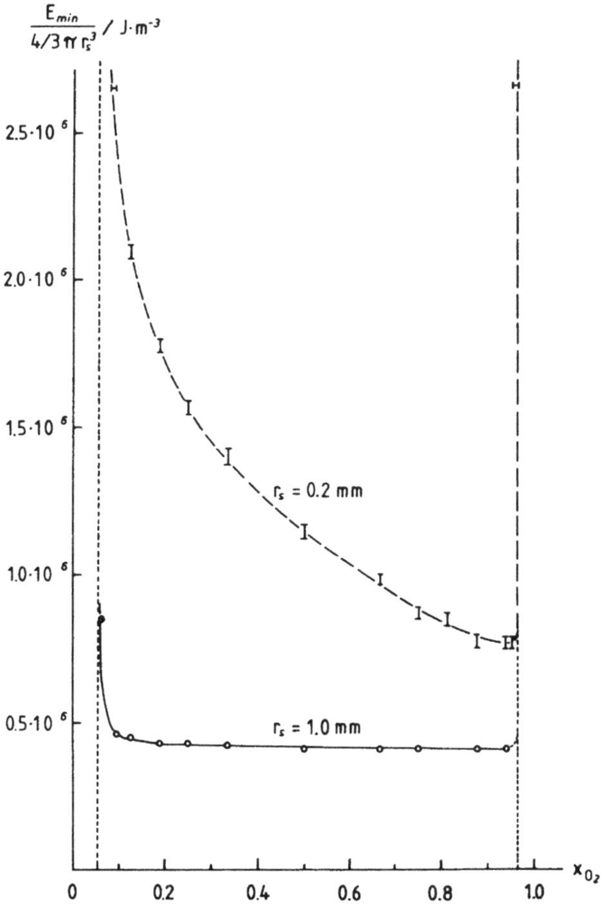

Fig. 10: Calculated minimum ignition energies in hydrogen-oxygen mixtures for different mixture compositions (uniform pressure assumption), spherical geometry, t_s = 0.1 ms, P = 1 bar

4.4 Results on Auto-Ignition in Hydrogen-Oxygen Mixtures

Since the detection of P-T explosion limits in H_2-O_2 mixtures in the thirties and forties, many efforts have been made to explain this phenomenon quantitatively. All of these attempts included severe simplifications as e.g. steady state assumptions, truncation of the reaction mechanism, or even restriction to homogeneous (zero-dimensional) systems.

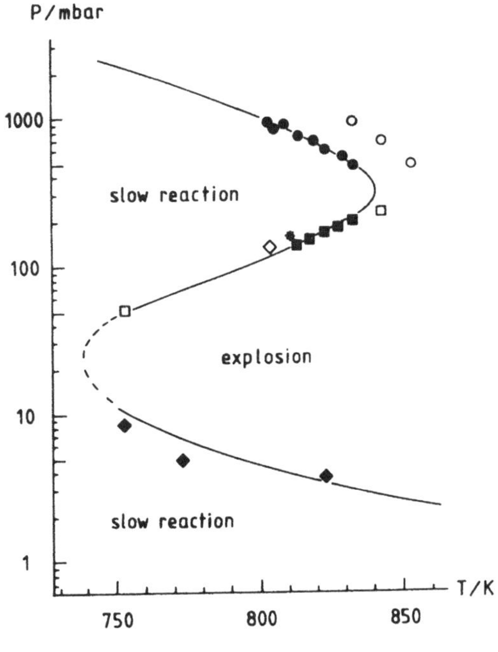

Fig. 11: Calculated (lines) and experimental (points) ignition limits for 2:1 mixtures of hydrogen and oxygen in vessels of different diameters; see [35] for further reference

The mathematical model described above is able to simulate all the explosion limits of the hydrogen-oxygen system without those simplifications, using detailed chemistry, a multi-species transport model and realistic surface chemistry. Ignition limits are determined by simulating the reaction in closed vessels (i.e. V = const.) for various initial pressures and temperatures.

At the explosion limits, there is a quite sharp transition from slow reaction (moderate temperature rise over a long time, sometimes more than 100 s) to ignition (with an immediate temperature rise in the center of the vessel and propagation of the flame).

Calculated P-T ignition limits for stoichiometric hydrogen-oxygen mixtures are shown in Fig. 11. As can be seen, calculated and experimental values are in quite a good agreement.

5. INSTATIONARY 2D PROBLEMS

As another example for some typical methods used to solve stiff problems, an instationary 2D reactive flow (full chemistry in terms of sets of elementary reactions) shall be considered.

Due to the large computational effort, only recently calculation of 2D Navier-Stokes equations including detailed chemistry in terms of sets of elementary reactions has become possible. An attempt for stationary cases has been reported [36]; the present paper demonstrates an extension to instationary cases.

5.1 Problem and Solution Method

Mathematical simulation of ignition processes have been performed by solving the corresponding system of conservation equations. The equation system [37] is simplified by using the ideal gas law and restricting the problem to 2D geometries (infinite rectangular column, finite cylinder). The finite cylinder geometry is considered here in detail; see [38,39] for further details.

The partial differential equation system, which consists of one algebraic (ideal gas law) and $n_s + 4$ partial differential equations (continuity, momentum, energy and species conservation equations; n_s = number of species), form an initial-boundary value problem for the dependent variables which are actually density ρ, velocities v_r and v_z, temperature T, and pressure P.

Again, spatial discretization of all derivatives different from that of the convective terms is performed by a standard central difference approximation for non-uniform grids, using nine-point stencils for the dissipative terms (to treat mixed second derivatives) and five-point stencils in all other cases.

The convective terms of the conservation equations have to be treated differently, because standard central difference approximations can cause severe numerical instabilities (e.g. overshoots) in regions of high gradients and curvatures. These instabilities can be avoided by using upwind schemes. However a major disadvantage of upwind schemes is the low order of approximation, leading to serious numerical diffusion.

In order to avoid this numerical diffusion, the approach of this work uses central difference approximations whenever possible and a modified upwind scheme whenever necessary (i.e. if overshoots might take place). In this case, locally monotonic interpolants were constructed as developed by *Fritsch and Butland* [33]).

One further problem arising in the numerical solution of the partial differential equation system considered is the resolution of shock fronts. The thickness of shocks is about 1 mm, and severe numerical instabilities result if the shock is not resolved by the mesh. Because it is not necessary to resolve the shock front for the problems considered in this work, an artificial viscosity term ("numerical diffusion") proposed by *Richtmyer and Morton* [34]) is applied, which spreads the shocks over a certain number of grid points.

The system of ordinary differential/algebraic equations (resulting from spatial discretization) again can be solved using the extrapolation code *LIMEX* [22,23]. This globally implicit method requires the evaluation of the Jacobian matrix. Using nine-point stencils for the spatial discretization (due to mixed second derivatives), the Jacobian has a block-nonadiagonal structure.

The dimension of the Jacobian is given by $n_{pde} \cdot n_l \cdot n_m$ where n_l and n_m are the numbers of grid points in the different directions and n_{pde} the number of partial differential equations. Even for the ozone-oxygen system considered, the dimension of the Jacobian is 16000 (using a 40 x 50 spatial mesh). The solution of the linear equation systems $A \cdot x = z$ (required by the implicit method) is performed by a block-iterative *Gauss-Seidel* method. The algorithm used for the iteration is given in more detail elsewhere [40].

The simulation of the ozone-oxygen ignition (see below) takes about 15 hours on an *IBM 3090*.

5.2 Results and Discussion

The model described above is tested simulating the 2D ignition process in the ozone-oxygen system in a cylindrical reaction vessel (28% O_3, P_0 = 0.34 bar). Thermal ignition occurs in a pocket with a diameter r_s = 1.37 mm and an ignition energy E_s = 10.9 mJ along the axis with decreasing energy density (in order to simulate absorption of energy). Corresponding experiments are described in [30,41]

Fig. 12 shows spatial profiles of temperature and pressure in the reaction vessel at different times, beginning at t = 1 μs, i.e. just after the external energy source has been turned off. Since the heating period is very short, there is no possibility for the pressure to equilibrate all over the reaction volume, and the pressure increase is approximately proportional to the temperature increase.

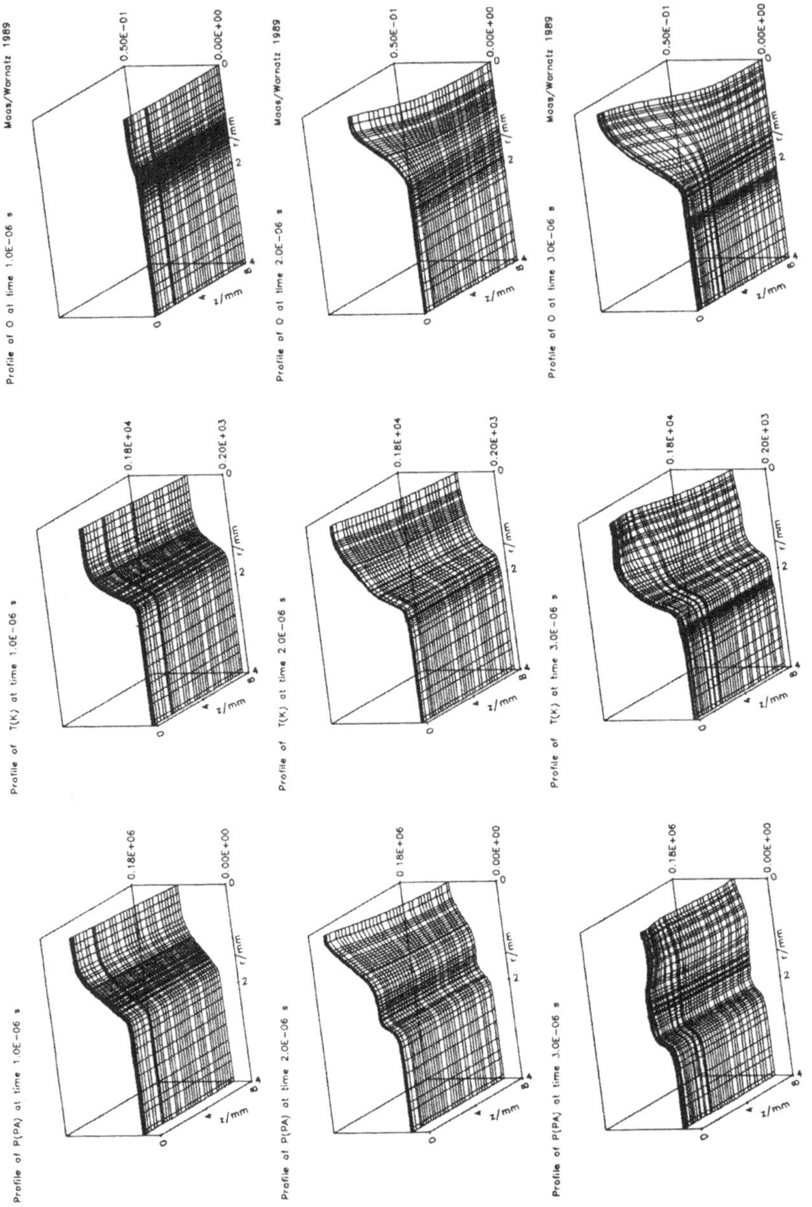

Fig. 12a: Calculated time dependent P, T, and O atom profiles; time = 1 μs - 3 μs (see text)

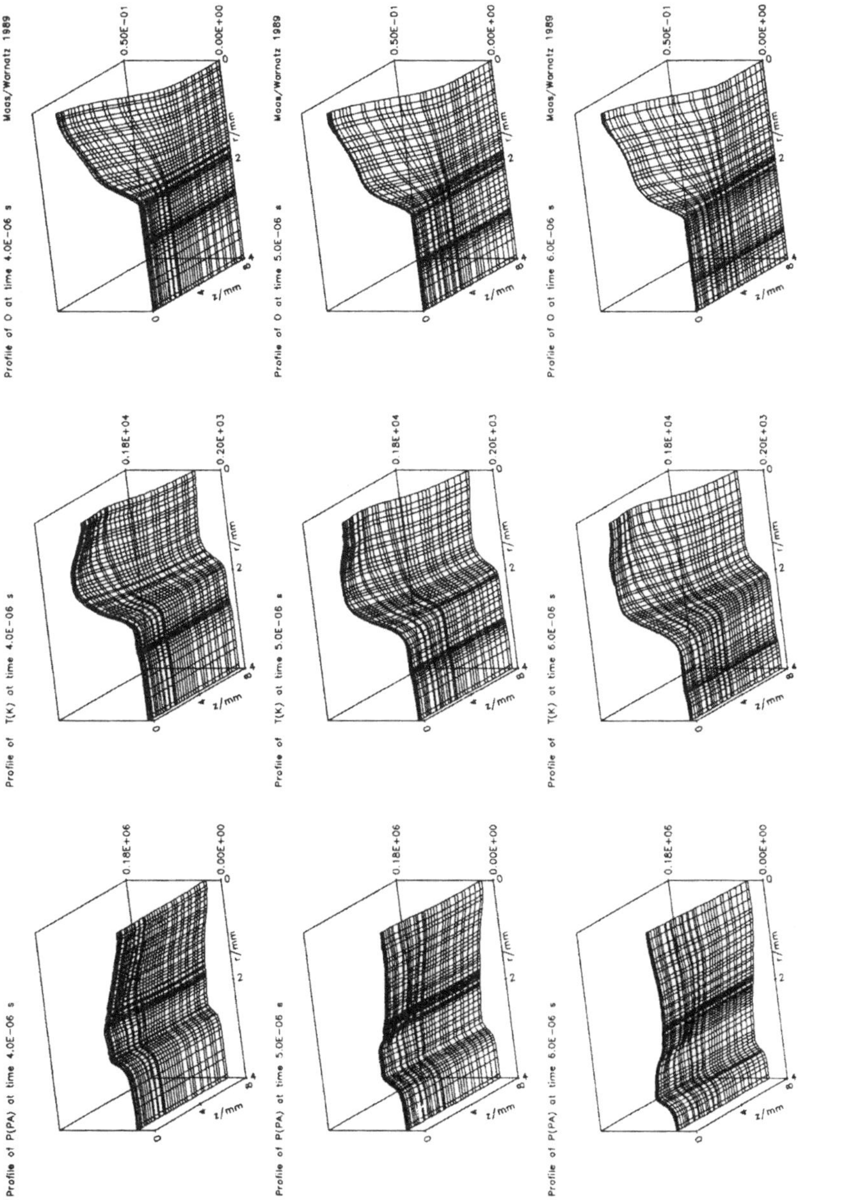

Fig. 12b: Calculated time dependent P, T, and O atom profiles; time = 4 μs - 6 μs (see text)

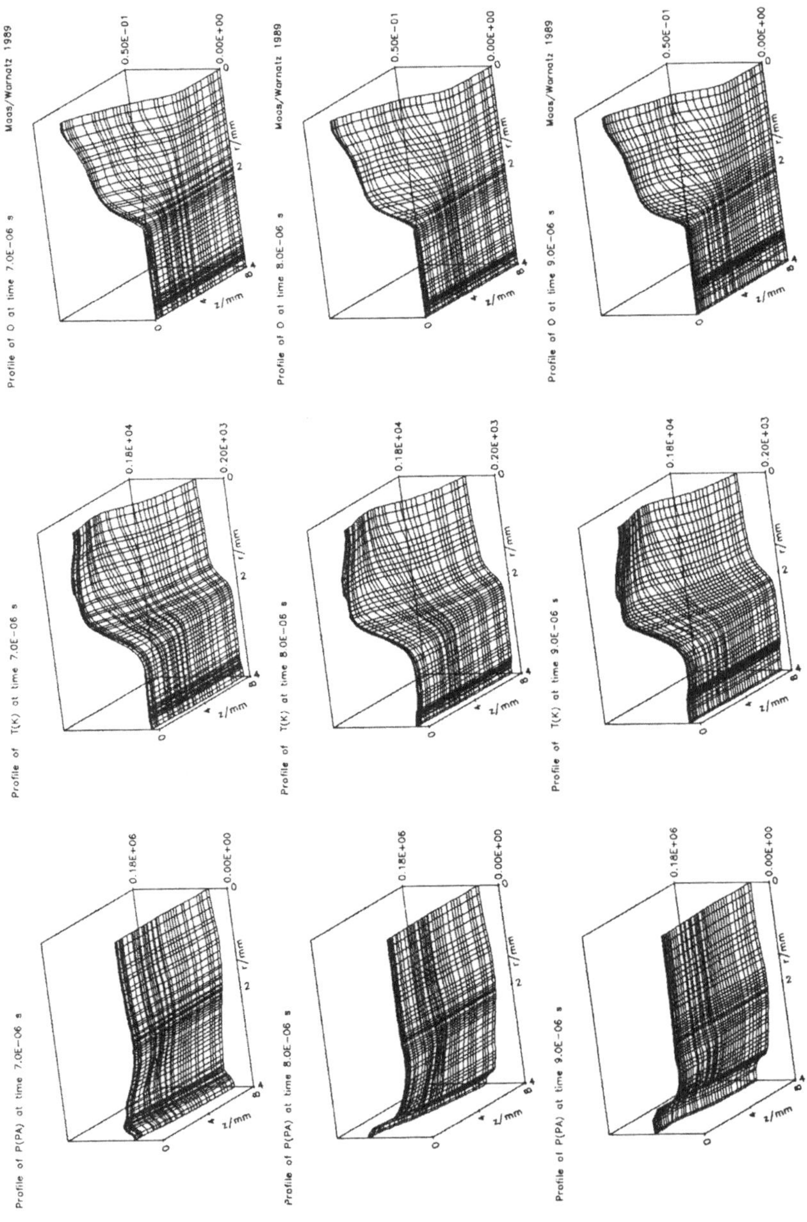

Fig. 12c: Calculated time dependent P, T, and O atom profiles; time = 7 μs - 9 μs (see text)

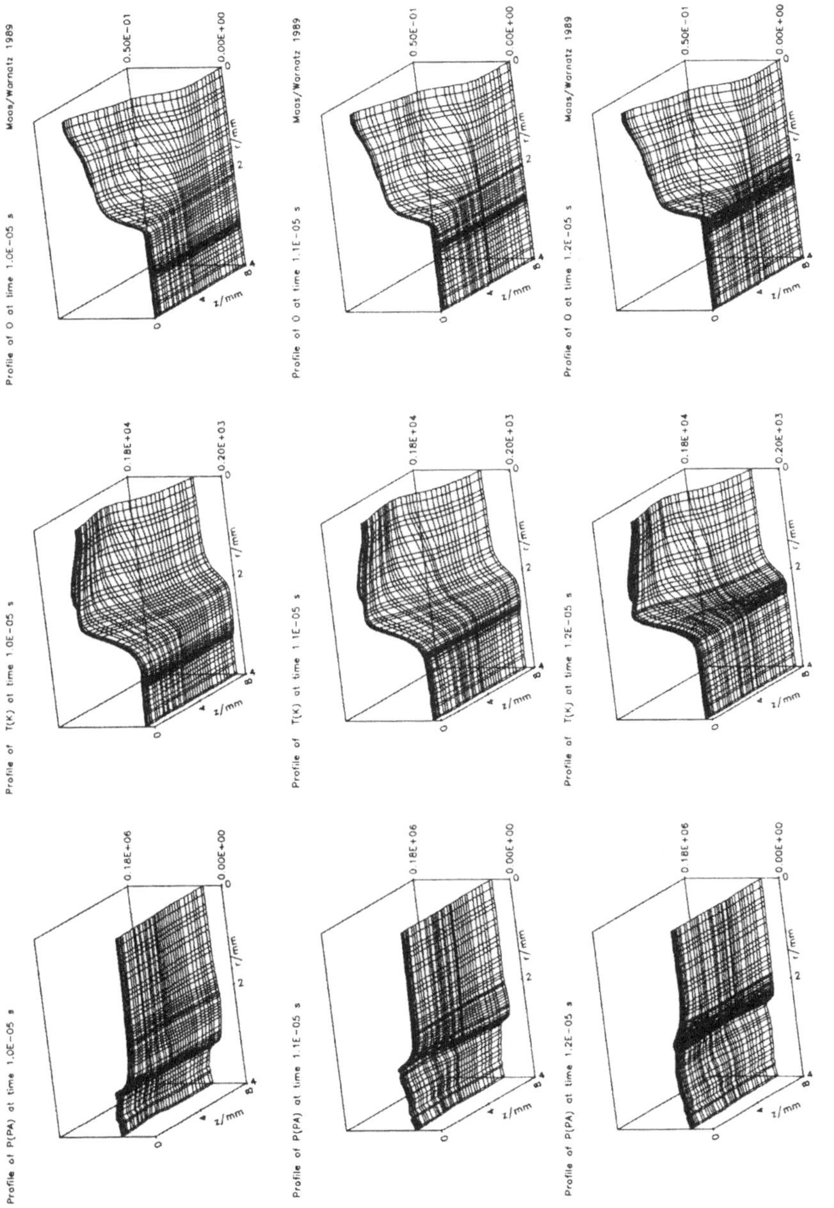

Fig. 12d: Calculated time dependent P, T, and O atom profiles; time = 10 μs - 12 μs (see text)

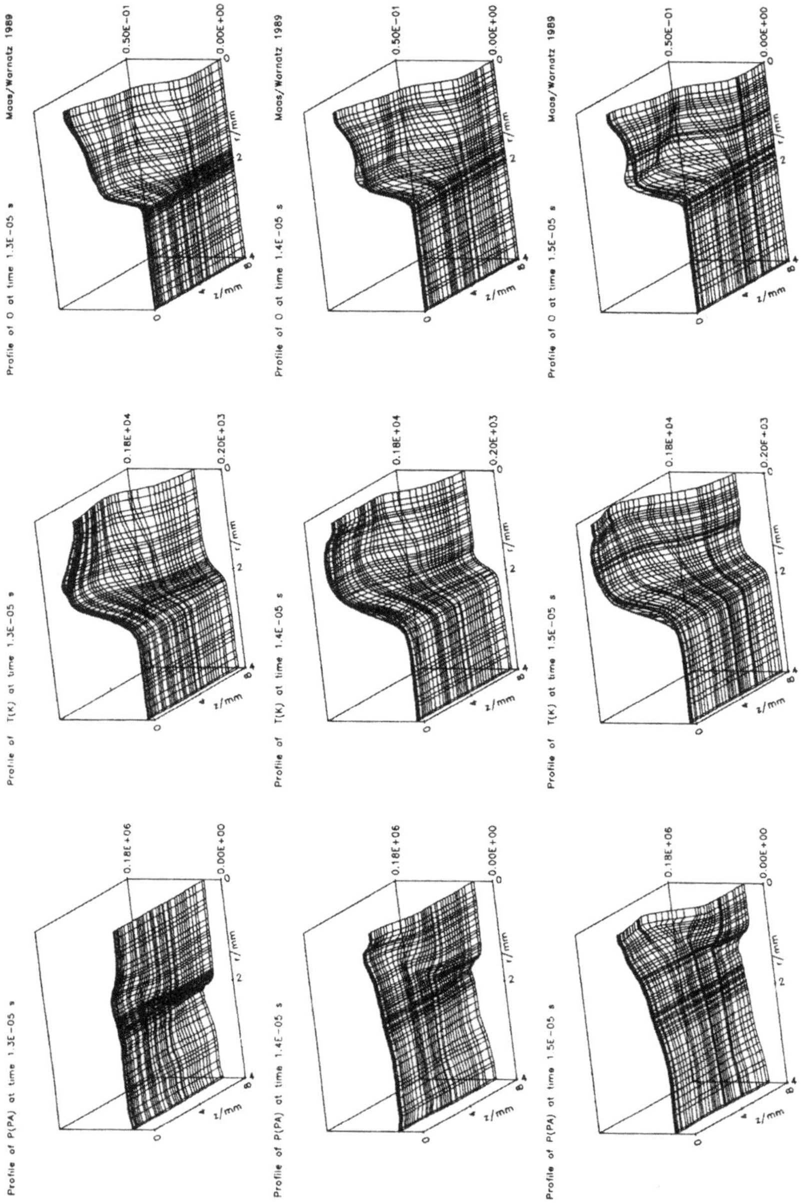

Fig. 12e: Calculated time dependent P, T, and O atom profiles; time = 13 μs - 15 μs (see text)

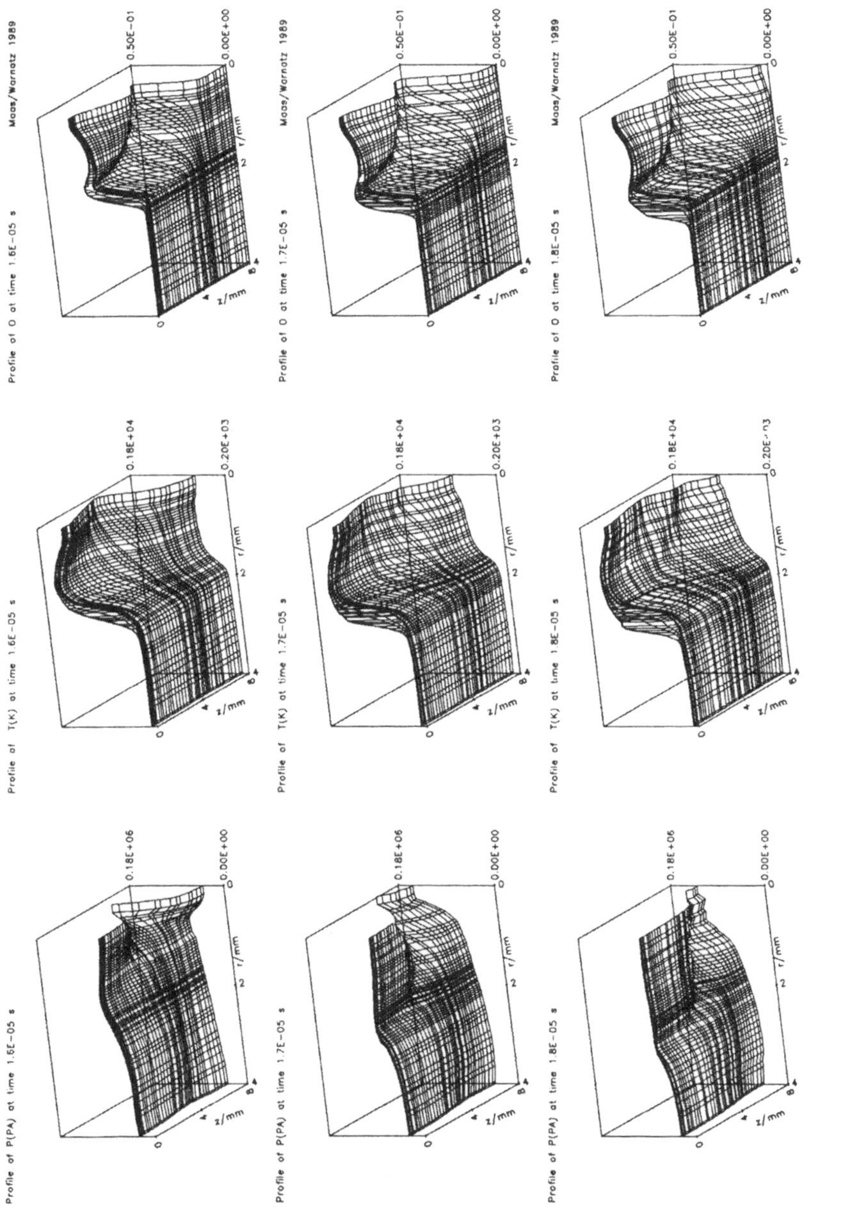

Fig. 12f: Calculated time dependent P, T, and O atom profiles; time = 16 μs - 18 μs (see text)

Due to the fact that the laser pulse used for ignition is diminished by absorption, flame propagation can be noticed both in radial and in axial direction in accordance with experimental results [30,41]).

Caused by the short ignition time (1ms in the example considered), a pressure wave is generated which is travelling to the wall, then reflected by the wall, and then interacting with the incoming reaction wave. In the vessel center, large pressure peaks can then be seen because of the focussing of the pressure wave.

6. ACKNOWLEDGEMENTS

The author is grateful to the "*Fonds der Chemischen Industrie*" and the "*Max-Buchner-Stiftung*" for financial support of this work and thanks *Dr. U. Maas* for valuable help.

7. REFERENCES

[1] *D. R. Stull, H. Prophet* (eds.): "JANAF Thermochemical Tables". U.S. Department of Commerce, Washington D.C. (1971), and addenda

[2] *A. Burcat*, in: W. C. Gardiner (ed.), "Combustion Chemistry". Springer, New York/Heidelberg (1984)

[3] *R. J. Kee, F. M. Rupley, J. A. Miller*: "The CHEMKIN Thermodynamic Data Base". SANDIA Report SAND87-8215 (1987)

[4] *J. O. Hirschfelder*: 9th Symposium (International) on Combustion, p. 553. Academic Press, New York (1963)

[5] *R. J. Kee, J. Warnatz, J. A. Miller*: "A Fortran Program Computer Code for the Evaluation of Gas-Phase Viscosities, Conductivities, and Diffusion Coefficients". SANDIA Report SAND83-8209 (1983)

[6] *R. J. Kee, G. Dixon-Lewis, J. Warnatz, M. E. Coltrin, J. A. Miller*: "A Fortran Computer Code for the Evaluation of Gas-Phase Multicomponent Transport Properties". SANDIA Report SAND86-8246 (1986)

[7] *R. Zellner*: "Bimolecular Reaction Rate Coefficients", in: W. C.

Gardiner jr. (ed.), Combustion Chemistry. Springer-Verlag, New York (1984)
[8] *W. C. Gardiner, J. Troe*: "Rate Coefficients of Thermal Dissociation, Isomerization, and Recombination Reactions", in: W. C. Gardiner jr. (ed.), Combustion Chemistry. Springer-Verlag, New York (1984)
[9] *D. L. Baulch, T. Just, J. A. Kerr, M. Pilling, J. Troe, R. W. Walker, J. Warnatz*: "Compilation of Rate Data on C1/C2 Species Oxidation". J. Phys. Chem. Ref. Data, publication in preparation
[10] *J. Warnatz*: Comb. Sci. Technol. 34, 177 (1983)
[11] *P. J. Robinson, K. A. Holbrook*: "Unimolecular Reactions". Wiley, London (1972)
[12] *B. Lewis, G. v. Elbe*: "Combustion, Flames, and Explosions in Gases". Academic Press, New York (1961)
[13] *J. Warnatz*: 18th Symposium (International) on Combustion, p. 369. The Combustion Institute, Pittsburgh (1981)
[14] *J. Warnatz*, in: W. C. Gardiner (ed.), "Combustion Chemistry". Springer, New York (1984)
[15] *J. Warnatz*: 20th Symposium (International) on Combustion, p. 845. The Combustion Institute, Pittsburgh (1984)
[16] *J. Warnatz, H. Bockhorn, A. Möser, H. W. Wenz*: 19th Symposium (International) on Combustion, p. 197. The Combustion Institute, Pittsburgh (1983)
[17] *J. Warnatz*: Ber. Bunsenges. Phys. Chem. 87, 1008 (1983)
[18] *F. Dryer, I. Glassman*, in: C. T. Bowman, J. Birkeland (eds.), "Alternative Hydrocarbon Fuels, Combustion and Kinetics". AIAA, New York (1979)
[19] *F. O. Rice*: JACS 55, 3035 (1933)
[20] *C. Esser, U. Maas, J. Warnatz*: Proc. Intl. Symp. on Diagnostics and Modelling of Combustion in Reciprocating Engines, p. 335. The Japanese Society of Mechanical Engineers, Tokyo (1985)
[21] *L. R. Petzold*: "A Description of DASSL: A Differential/Algebraic System Solver", Report SAND82-8637. Sandia National Laboratories, Livermore (1982); IMACS World Congress, Montreal 1982
[22] *P. Deuflhard, E. Hairer, J. Zugck*: "One-Step and Extrapolation Methods for Differential/Algebraic Systems". Univ. Heidelberg, SFB 123: Tech. Rep. 318, (1985)
[23] *P. Deuflhard, U. Nowak*: "Extrapolation Integrators for Quasilinear Implicit ODEs". Univ. Heidelberg, SFB 123: Tech. Rep. 332, (1985)
[24] *J. R. Leis, M. A. Kramer*: Computers and Chemical Engineering 9, 93-96 (1985)

[25] *M. Caracotsius, W. E. Stewart*: Computers and Chemical Engineering 9, 359-365 (1985)
[26] *W. J. Pitz, J. Warnatz, C. K. Westbrook*: 22nd Symposium (International) on Combustion, p. 893. The Combustion Institute, Pittsburgh (1989)
[27] *R. Zellner, K. Niemitz*: publication in preparation
[28] *P. Pinchon*: personal communication
[29] *B. Raffel, J. Warnatz, H. Wolff, J. Wolfrum, R. J. Kee*, in: J. R. Bowen, J.-C. Leyer, R. I. Soloukhin (Eds.), "Dynamics of Reactive Systems", Part II, p. 335. AIAA, New York (1986)
[30] *U. Maas, B. Raffel, J. Warnatz, J. Wolfrum*: 21st Symposium (International) on Combustion, p. 1869. The Combustion Institute, Pittsburgh (1987)
[31] *B. Raffel, J. Warnatz, J. Wolfrum*: Appl. Phys. B37, 189 (1985)
[32] *U. Maas, J. Warnatz*, in: A. L. Kuhl, J. R. Bowen, J.-C. Leyer, A. Borisov (eds.), "Dynamics of Reactive Systems", Part I, p. 3. AIAA, New York (1988)
[33] *F. N. Fritsch, J. Butland*: SIAM J. Sci. Stat. Comput. 5, 300 (1984)
[34] *R. Richtmyer, K. Morton*, in: L. Bers, R. Courant, J. Stoker (eds.), "Interscience Tracts in Pure and Applied Mathematics" No. 4, 2nd edition (1967)
[35] *U. Maas, J. Warnatz*: Comb. Flame 74, 53 (1988)
[36] *M. D. Smooke*: 3rd Workshop on Numerical Problems in Flame Propagation. Leeds (1988), publication in preparation
[37] *R. B. Bird, W. E. Stewart, E. N. Lightfoot*: "Transport Phenomena". John Wiley & Sons, New York (1960)
[38] *U. Maas, J. Warnatz*: Zeitschrift für Physikalische Chemie NF 161, 61 (1988)
[39] *U. Maas, J. Warnatz*: Impact of Computing on Science and Engineering (1989), submitted
[40] *D. Marsal*: "Die numerische Lösung partieller Differentialgleichungen in Wissenschaft und Technik". Bibliographisches Institut Mannheim (1976)
[41] *B. Raffel, J. Wolfrum*: Zeitschrift für Physikalische Chemie NF 161, 43 (1988)

TURBULENT COMBUSTION

F. A. Williams
Professor of Engineering Physics and Combustion
Department of Applied Mechanics and Engineering Sciences
University of California, San Diego
La Jolla, CA 92093

1. FUNDAMENTALS

Turbulent combustion is fundamentally based on the conservation equations of chemically reacting flows. These equations are composed of the set of partial differential equations of fluid mechanics, augmented by the equations of chemical kinetics for multicomponent systems. The equations are summarized at the beginning of Ref. 1. Review of the set of equations is relevant since in addressing turbulent combustion often authors implicitly neglect effects by beginning with simplified conservation equations.

The conservation equations may be derived phenomenologically from continuum theory or deductively from the kinetic theory of dilute gases. Both approaches are useful, and they are complementary. There are particular problems in which further complications may arise from the necessity of considering coupling with the radiation transport equation, with equations of electrodynamics, with equations of multiphase flow, etc. However, attention is restricted here to turbulence in reacting ideal gas mixtures at low enough radiation and ionization levels that the cited underlying equations are sufficient.

Many of the approaches to the theory of turbulent combustion can be viewed as being based on these reacting Navier-Stokes equations. A list of kinds of approaches to the subject is given in Table 1. Simulation by direct numerical integration will evolve as computers progress. Approaches based on evolution of probability-density functions (PDF's) have advanced with computational capabilities. Approximation of PDF's by assumed functional forms, with evolution then determined by classical moment methods, often has been found quite useful. Age theories, like the ESCIMO approach of Spalding, represent those not directly based on the reacting Navier-Stokes equations. Perturbation methods have been restricted mainly to large-scale, low-intensity turbulence. More detailed discussions of these approaches are given in Ref. 2.

The last two categories listed in Table 1 are too new to have been addressed in Refs. 1 and 2. Recent issues of *Combustion Science and Technology* may be consulted for papers on this. Methods of the renormalization group have been applied recently by Yakhot and by Sivashinsky and have been discussed by Kerstein in this journal. They begin at small scales or low turbulence intensities and from perturbation results derive expressions such as formulas for turbulent burning velocities applicable at large scales and high intensities. Fractals are geometrical properties that can be convenient for data reduction but that are consequences rather than drivers of the dynamics of turbulent combustion. These represent active areas of research from which more may be expected in the future.

It is convenient to identify regimes of turbulent combustion in a plane of Damköhler number as a function of Reynolds number. Such a graph is shown in Fig. 1. Here ℓ is the integral scale of turbulence, k the turbulent kinetic energy, ν the kinematic viscosity, and τ_c a characteristic chemical time. The turbulence Reynolds number based on the integral scale is denoted by $R_\ell = \ell\sqrt{2k}/\nu$, and the Damköhler number based on the integral time is denoted by $D_\ell = \left(\ell\sqrt{2k}\right)/\tau_c$. These have been selected as the coordinates for the graph. The Reynolds number based on the Taylor scale is $R_t = \sqrt{R_\ell}$, and the Damköhler number based on the Kolmogorov time τ_k is $D_k = \tau_k/\tau_c$. Here the characteristic time scale of Kolmogorov eddies is $\tau_k = \sqrt{\nu/\varepsilon}$, where the average rate of dissipation of turbulent kinetic energy has been defined as $\varepsilon = (2k)^{3/2}/\ell$. In constructing this graph, for simplicity and clarity all multiplicative constants have been put equal to unity. All lines should therefore be considered fuzzy by an amount of perhaps an order of magnitude, but because of the wide range of values covered by the graph, this fuzziness does not materially affect the general conclusions.

The advantage of Fig. 1 over similar graphs shown previously is that it applies for both premixed flames and diffusion flames. Most earlier maps of this kind have been considered only for premixed flames. In such cases $\sqrt{\nu/\tau_c}$ is a laminar burning velocity S_L, τ_c being a suitable transit time through a premixed laminar flame, and $\sqrt{\nu\tau_c}$ is a representative laminar-flame thickness δ_L. Since diffusion flames cannot be characterized by an intrinsic propagation velocity S_L or an intrinsic thickness δ_L, only τ_c remains for use in Fig. 1 for diffusion flames. For premixed flames Fig. 1 is entirely equivalent to the familiar Barrère-Borghi diagram, and also to a diagram proposed by Bray in Ref. 3, but in both cases with skewed coordinates.

There are two principal limiting regimes in Fig. 1, the distributed-reaction regime at small D_ℓ and the reaction-sheet regime at large D_ℓ. In the former all turbulence scales are short compared with chemical scales, and combustion occurs in a distributed manner throughout the turbulent flow. In the latter all turbulence scales are long compared with chemical scales and the chemistry occurs in thin sheets moved and wrinkled by the turbulence. A sufficient condition for the former regime is $l/\sqrt{\nu\tau_c} << 1$, and a sufficient condition for the latter is $D_k >> 1$. From Fig. 1 it is seen that at large R_ℓ there are intermediate regimes falling not entirely in either regime. Although these intermediate regimes have not been named, the region between $D_k = 1$ and $D_\ell = 1$ might be called the broken-flamelet regime, since reaction sheets may be anticipated in large eddies but not in small eddies in this regime.

For premixed flames the line $S_L = \sqrt{2k}$ has been suggested as a rough boundary between single and multiple reaction sheets. This line thus divides the reaction-sheet regime into two subregimes. The upper left-hand corner is a limit of the single-sheet subregime in which the turbulence perturbs the laminar flame only slightly. This is indicated as the weak-turbulence limit in Fig. 1. The importance of Fig. 1 stems from the fact that any one of the approaches of Table 1 is better adapted to one regime than to another. Before selecting a method of analysis, every effort should be made to identify as closely as possible the regime of turbulent combustion in the problem.

2. TURBULENT DIFFUSION FLAMES

Turbulent diffusion flames in the distributed-reaction regime are expected to resemble premixed turbulent flames in that regime, in that the rapid turbulent mixing will cause the chemistry to occur in a more or less premixed fashion – the differences reside mainly in the boundary conditions. Because of the slow chemistry, fluctuations in temperature and in species concentrations are expected to be small, and eddy diffusivities can be employed to characterize turbulent transport with reasonable accuracy. Chemical production rates can be expressed as the rates evaluated at local average thermodynamic conditions plus small corrections. As a consequence of these simplifications, classical moment methods are well suited for describing turbulent combustion in this regime. Methods of evolution of PDF's also can be used here if information concerning the PDF's is desired.

Turbulent diffusion flames in the reaction-sheet regime can be analyzed conveniently by the method of approximation of PDF's, in conjunction with moment methods. In the reaction-sheet regime, the fluctuations of temperature and species compositions are large, and therefore classical moment methods applied to flow variables with chemical source terms are inaccurate. The difficulty extends into turbulent transport in that, in such cases, it has been found that countergradient diffusion occurs – when a diffusion approximation is employed for turbulent transport the diffusion coefficient changes sign within the flow. Because of these difficulties it is inappropriate to apply moment methods to quantities with chemical source terms in this regime. However, moment methods may still be employed for conserved scalars or the so-called coupling functions, that is, variables (like total enthalpy) that do not have chemical source terms. The mixture fraction is a convenient normalized conserved scalar. The approach is to apply moment methods to the mixture fraction, to approximate its PDF (e.g., through beta functions or clipped Gaussians), to employ the results of the moment methods to calculate the parameters describing this PDF, and then to use laminar reaction-sheet results, that is, so-called flamelet structures, to obtain PDF's for temperature and species compositions from the mixture-fraction PDF. The procedure is described in detail in Ref. 1 and is reviewed further in Ref. 2.

An interesting application of this method concerns turbulent-jet diffusion flames. Their general characteristics are illustrated in Fig. 2. Liftoff involves extinction of flamelets in the lower part of the flame where the strain rates are largest. In Fig. 1 this corresponds to crossing the line $D_k = 1$. Towards the end of the flame, at the flame tip, the reaction-sheet regime should apply until blowoff. However, this idea has been challenged recently through claims that there are fluctuations within the reaction sheets. This is a point of controversy concerning which continuing research may be anticipated. Basic aspects of turbulent-jet diffusion flames are presented in Refs. 1, 2 and 3.

3. PREMIXED-FLAME INSTABILITIES

Behaviors of premixed turbulent flames may be expected to be influenced by instabilities of premixed laminar flames. There are many physical mechanisms for combustion instabilities, as indicated in Ref. 1. Three that are particularly relevant to premixed laminar flames are body-force instabilities, hydrodynamic instabilities, and diffusive-thermal instabilities. These have been listed here in the order of decreasing scale. The basics of all of these three types of instability are presented fully in Ref. 1, which provides literature citations to other important review articles, such as that of

Clavin. From Clavin's work it is known that the hydrodynamic instability, which always is destabilizing, can be stabilized fully only by a combination of two other phenomena, such as stabilizing diffusive-thermal effects at small scales and stabilizing body-force or wall effects at large scales. This information may be found in Ref. 1.

4. TURBULENT PREMIXED FLAMES

The regimes of premixed turbulent flame propagation can be considered on the basis of Fig. 1, as previously indicated. Velocity and length scales now become relevant, so that further significance can be ascribed to the various lines, as has been explained. In the distributed-reaction regime, the moment methods or methods of PDF evolution can be applied, as has been described for turbulent diffusion flames. Eddy breakup models may be viewed as attempts to extend these methods into reaction-sheet regimes. These methods have encountered successes for certain problems and are described in Ref. 1. In the distributed-reaction regime, it is quite questionable whether there exists a turbulent flame velocity that is a function only of the thermodynamic and turbulence properties of the unreacted gas. This is another aspect in which premixed turbulent flames in the distributed-reaction regime need to be treated differently from those in the reaction-sheet regime.

In the reaction-sheet regime the dynamics of wrinkled laminar flames become relevant to premixed turbulent flame propagation. Discussions of these dynamics are given in Ref. 1 and in the review by Clavin. It is possible to define flame stretch in a general way to include effects of both strain rate and flame curvature when the stretch is weak. The propagation velocity of the laminar flame then depends on the stretch through a coefficient represented as the Markstein length or, nondimensionally, as the Markstein number, b. Evolution equations for the flame can then be derived. The dynamics of wrinkled flames are conveniently studied on the basis of these evolution equations. References 1 and 2 summarize the procedures. More recently a field-equation approach has been found useful, as summarized in Fig. 3.

Turbulent burning velocities for premixed flames in the reaction-sheet regime may be addressed by a number of different approaches. Practically all of the approaches in Table 1 have been applied. Some results are listed in Table 2.

A fundamental principle is that in the reaction-sheet regime the ratio of turbulent to laminar burning velocities is independent of the scale of turbulence. Arguments to the contrary can be given, but these arguments are not definitive. The flame-speed formulas quoted in Table 2 thus express the ratio as a function of the ratio of the turbulence intensity u' to the laminar burning velocity. Formula 2 is obtained from perturbation methods and is good at low turbulence intensities. Formula 5 was calculated from an inventive application of the PDF evolution method but unlike other formulas, with the exception of formula 3, exhibits no influence of laminar flame propagation at high turbulence intensities. In principle some such dependence should remain, and it has disappeared here because of the turbulent-diffusion approximation in the PDF-evolution method.

Formulas 6 and 7 come from renormalization methods recently published in *Combustion Science and Technology*. Formula 6 is especially attractive, although its derivation is somewhat unclear. This formula reduces to formula 2 at low turbulence intensities and gives a result at high turbulence intensities in which the linear proportionality between S_T and u' is corrected only by a logarithmic factor, a minimal correction including effects of flame propagation on turbulent dispersion. The first five formulas are discussed in Refs. 1 and 2. Formulas 6 and 7, the latter exhibiting influences of the type of turbulence on the burning velocity, are interesting new topics for continuing study.

Experimentally formula 3 has been obtained, with results similar to formula 4. However, experimental measurement of S_T is notoriously difficult. An important item of future research is the careful evaluation of S_T data and the design and completion of better experiments. There is much more to be done in the field of premixed turbulent flame propagation.

REFERENCES

1. Williams, F.A., *Combustion Theory*, Second Edition, Benjamin/Cummings, Menlo Park, CA (1985).
2. Williams, F.A., "Turbulent Combustion," Chapter III of *The Mathematics of Combustion* (J.D. Buckmaster, editor), Society for Industrial and Applied Mathematics, Philadelphia, PA, 1985, pp. 97-131.
3. Libby, P.A. and Williams, F.A., editors, *Turbulent Reacting Flows*, Vol. 44 of Topics in Applied Physics, Springer-Verlag, Berlin Heidelberg New York (1980).

Table 1.

Approaches to Theories of Turbulent Combustion

1. Direct Numerical Integration
2. Evolution of PDF's
3. Approximation of PDF's
4. Moment Methods
5. Age Theories, ESCIMO, and the Like
6. Perturbation Methods
7. Renormalization Methods
8. Other Methods (Fractals, etc.)

Table 2.

Formulas for Turbulent Burning Velocity

1. $S_T/S_L = f(u'/S_L)$ Independent of Scale $(bK \rightarrow 0)$

2. $S_T = S_L\left\{\frac{1}{2}\left[1+\sqrt{1+8c\left(\frac{u'}{S_L}\right)^2}\;\right]\right\}^{1/2}$ (Clavin) $(c = 1, c = 10)$

3. $S_T = S_L + c'u'$ (Empirical) $(c' = 1, c' = 4)$

4. $S_T = S_L[1 + 3.5(u'/S_L)^{0.7}]$ (Klimov) (Dispersion + Propagation)

5. $S_T = S_L + 2.1u'$ (Pope) (PDF Evolution)

6. $S_T = u'/[\ell n(S_T/S_L)]^{1/2}$ (Yakhot) (Renormalization)

7. $S_T = u'/g(S_T/S_L)$ (Sivashinsky) (Renormalization)

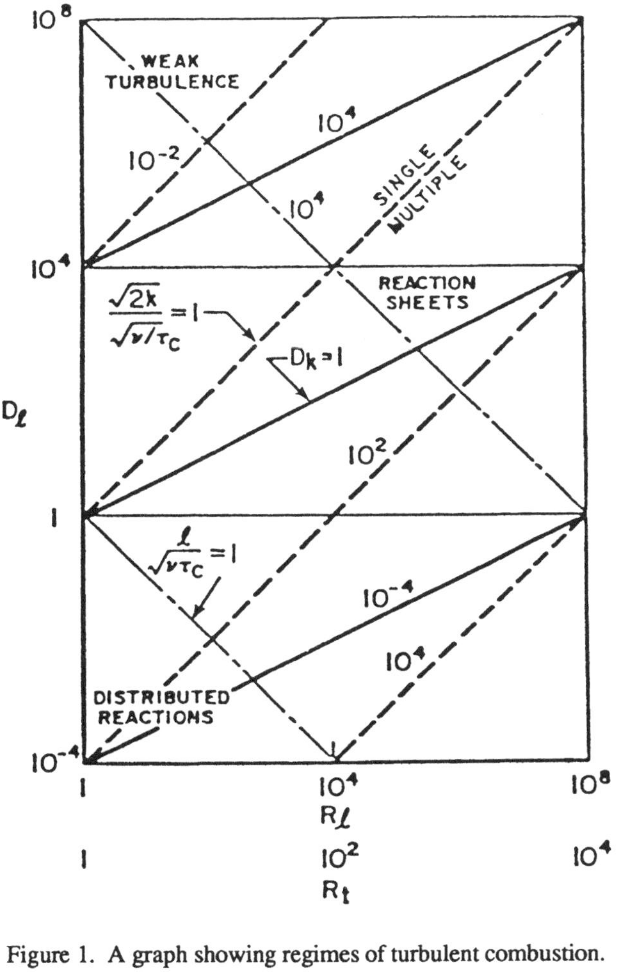

Figure 1. A graph showing regimes of turbulent combustion.

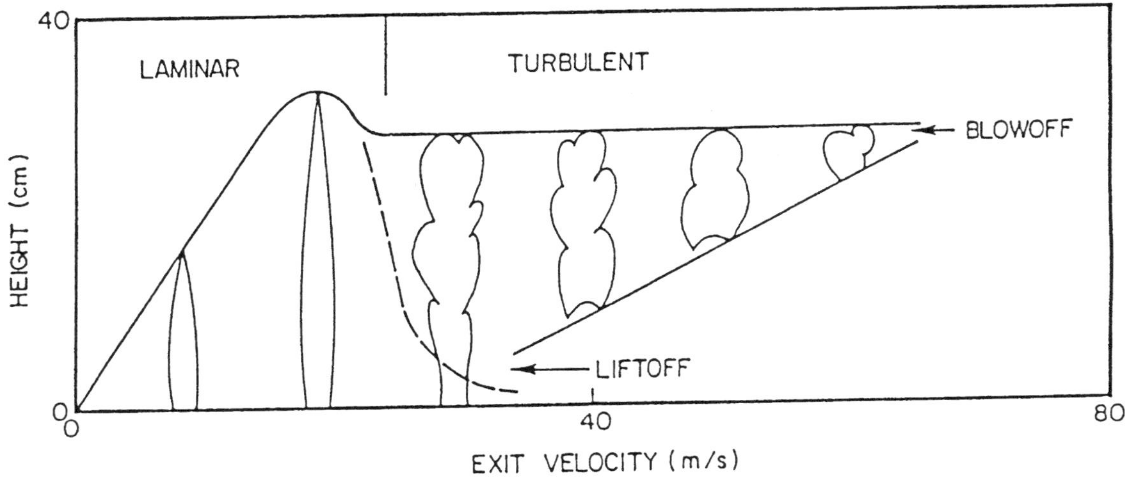

Figure 2. Schematic illustration of turbulent-jet diffusion flame.

Premixed Turbulent Flames
In the Reaction-Sheet Regime

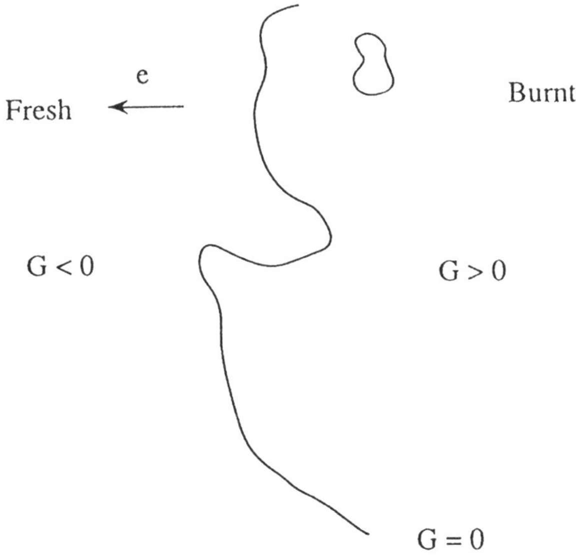

$$\frac{\partial G}{\partial t} + \mathrm{v} \cdot \nabla G = V|\nabla G|$$

$V = S_L(1 - bK)$ (Large Scale)

Nondimensional Flame Stretch

$$K = \frac{\delta_L}{S_L}\left\{S_L \nabla \cdot \mathrm{n} - \frac{1}{2}\mathrm{n} \cdot \left[(\nabla \mathrm{v}) + (\nabla \mathrm{v})^T\right] \cdot \mathrm{n}\right\}$$

Flame Curvature   Strain-Rate Tensor

Figure 3. Field-equation approach for describing premixed turbulent-flame propagation in the reaction-sheet regime.